Introduction to PLCs

Elvin Pérez Adrover

ISBN: 061565438X
ISBN-13: 978-0615654386

DEDICATION

To Elvin jr, and Omar.

CONTENTS

1 INTRODUCTION: RELAY LOGIC, LADDER DIAGRAMS, AND THE PROGRAMMABLE LOGIC CONTROLLER

In many instances in our daily living, we often find ourselves in the situation where there is an **action** taken depending on a **condition** happening. For example, we carry and open an umbrella or wear a raincoat (the action) if it happens to rain (the condition). We stop the car (the action) (for those of us who are not into traffic violations) if the traffic light happens to be red (the condition). Likewise, if we find a device that will perform an action based on a condition, we can arrange for some of these daily decisions to be carried out an autonomous way by using such a device. This is the principle of making things to happen or perform by themselves, or plainly speaking, **automating** them.

One of the simplest devices designed to execute such an action as described above is the **electromechanical relay** (refer to figure). On its most elementary sense, it consists of an electromagnet (the **coil**), which when energized (magnetized), pulls an actuator mechanically fixed to one or more pairs of plates (the **contacts**). An electrical current can be connected to these contacts, and they can, based on the electrical (magnetic) state of the coil, either come together (**close**, thus allowing the flow of the electrical current connected to them) or separate (**open**, then interrupting this flow of electrical current). Now if we either apply or interrupt an electrical current flow to the coil (the condition), an electrical circuit connected to a closing contact will energize or another connected to an opening contact will de-energize (the action). This is the basis of industrial automation, where we have a heavy need of performing diverse actions depending on equally diverse conditions.

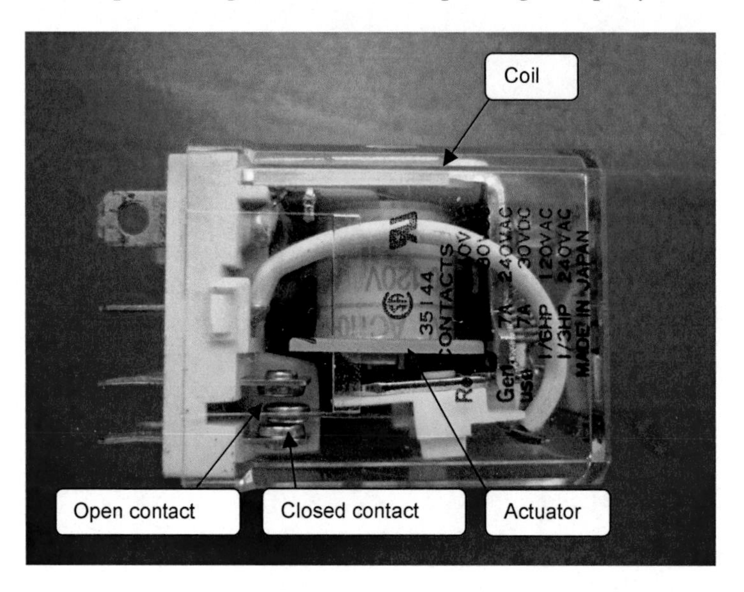

We can now make this scenario more complex if we go ahead and arrange for more than one contact to be connected into a single circuit. These contacts can be connected in series connection (thus arranging a logical **AND** of the conditions: this means that **all** of the conditions must be complied with in order for the circuit to be energized and the action to be performed). They can also be connected in a parallel fashion (thus arranging a logical **OR** of the conditions: this means that now **any** of the conditions must be satisfied for the circuit to be energized and the action performed). They can also be connected in combinations of the abovementioned arrangements, so the complexity of the circuit can now meet more and more stringent specifications.

The usual presentation for a relay is to have some kind of a base. Here the actual relay assembly can be plugged in. There, internal connections join plugs coming from either the relay contacts or the coil to connections that can be wired (usually with screw or spring terminals). In this way, wiring circuits from one relay base to another can be accomplished and these serial or parallel or combination of arrangements can be done.

For many years, these arrangements of electromechanical relays wired together were the heart and base of Industrial Automation. There were some variants, like **timing relays.** Here the applied current (condition) does not energize a coil, but rather energizes a mechanism that counts the time elapsed. After this preset time elapses, then associated contacts either open or close (action). There are also counters, where the application of current (condition) triggers a mechanism that counts or accumulates. After a preset quantity of counts, then the associated contacts are actuated (action).

The widely accepted notation to document such Automation circuits represents each circuit by means of a horizontal line, interrupted by the connected contacts, either normally open or closed. Normal here means the state of the contact when the coil is not energized. The horizontal line finishes in the coil (which is represented at the far right end). The voltage source (Line Voltage and Neutral if Alternating Current –AC, normal household current- or Positive and Negative if Direct Current –DC, typical current of a battery-) is represented by vertical lines. These are drawn along the left (Line or +) and right (Neutral or -) edges of the depiction. When correctly documented, each horizontal line can become a physical guide on how to wire a particular circuit. This arrangement of a pair of vertical lines joined by many horizontal ones resembles a ladder (see figure), so this type of diagram is called a **ladder diagram**.

Example of a ladder diagram

The Programmable Logic Controller (PLC)

These relay arrangements presented many difficulties in some aspects. Such a system is somewhat hard to design, as the designer must keep track of the contacts used for each relay in the design. This can be fairly simple if the project consists of maybe less than five or six relays, but as the project gets bigger in size, so grows its complexity in order to keep track of every contact used. On the other hand, wiring such a complicated scheme can be very hard as well, and will require a lot of Control Panel (enclosure or cabinet) space. The quantity of wires involved makes the process more prone to errors as it become more complex. Last, but not less important, making a change in the logic of such a system is cumbersome. Wiring to a different kind of contact means physically disconnecting and reconnecting many single wires at new locations, let alone the fact that, if we want to switch from a normally open to a normally close contact or vice-versa, we can run into the possibility of not having more available contacts of the sort. The situation of having to perform a change in a control system logic is fairly common in Industrial environments.

Let's say now that we can have a special computer whose memory can store all of this contact and wiring information. Now, our circuits and most of its associated wiring become virtual, where the contacts doesn't physically exist anymore, but are rather stored in this computer memory. Changing the arrangement of these contacts is now a function of editing this computer memory. And, as the contacts are virtual, there is no limit on the quantity of contacts that can be related to a single coil. Also, as we are dealing with computer memory, mathematical operations that can be executed bring some additional flexibility to our logic scheme not possible to dream of in a relay-based system. This wizard computer is the **Programmable Logic Controller (PLC)**.

Inputs and Outputs

Even though all the contacts in the relay logic for our PLC are virtual, we still need a mechanism to communicate and interact with the outside world. We do still need for something physical to either energize or de-energize, triggering a condition, and also need another something physical to be energized, thus creating the action. These "outside connections" to our PLC are the **Input/Output (I/O)** system related to our PLC.

When we define an "Input" or an "Output", we adopt the PLC's point of view: an **Input** stands for a signal that comes from the outside world to the PLC system. Conversely, and **Output** will be a signal that goes from the PLC system to the outside world. We can now identify the condition in our action/condition pair as coming from an Input, as well as the action as going through an Output.

There are mainly two types of Inputs and Outputs. The first one is described by a condition or signal that can be turned ON and OFF, and these two are its only states, for example, a light switch. A light switch can only be turned ON, allowing electrical flow to the light, thus energizing it, or OFF, interrupting the electrical flow and de-energizing the light. This can also be understood as a signal indicating either presence or absence of something. For example, a shower water heater works in the principle of energizing the heater element ("the resistance")(the action) when there is enough water pressure on the shower (the condition) and conversely de-energizing the element when there is not enough water pressure (as when we turn off the water). The detection of this presence/absence of enough pressure is performed by a **pressure switch**, which is a good example of this type of Input. The switch closes an electrical contact when the presence of this pressure is detected and vice-versa. It can only have two states: enough pressure, or not enough pressure.

These two states can be represented by a binary number system, where the only possible digits are 0 or 1. Then every Input of this sort will be represented by a **bi**nary digi**t** ("**bit**"), which takes the value of 0 when the Input is de-energized and the value of 1 when it is energized. As it can be represented by a binary digit, this type of Input is called **Digital**, or sometimes **Discrete**.

There is another possibility, where we need to represent an actual quantity that can vary over time, for example a temperature measurement. This is not a signal that can be represented as a presence or absence of a condition, but rather needs to be represented by a number proportional to the measured property, say 100°C or 25 psig. This kind of data is called **Analog**. Analog data is stored in the PLC memory, rather than by a single bit, by a group of bits called a **word**.

Returning to the outside connections to and from our PLC, it is important to mention that for most systems, these I/O are connected using modules. There are **Analog Input** and **Analog Output** modules, as well as **Digital Input** and **Digital Output** modules. Other types of I/O modules for special functions can be found, which are beyond the scope of our discussion here. There is also the possibility of concentrating I/O systems and connecting them to the PLC by means of dedicated communication networks. These will be discussed later on as a special topic. Even though smaller systems tend to have their I/O fixed in the same physical structure as the PLC CPU, or computer, or **Processor**, many of them have also the possibility of adding additional I/O modules. For systems based on modules, we often find a physical cage with prefabricated connections for these modules, called **rack** or **chassis**. Then, the Processor is another module to be placed in the chassis, along with the selected I/O. The fact that we can choose the I/O modules to be placed in our PLC system is the cornerstone of the PLC **modularity**, which makes the PLC suitable and flexible enough to accomplish many (almost infinite) process control tasks and combinations. When we design a PLC system, we decide how large it will be and what will be its exact composition. Also, there is always the possibility of either altering the established configuration at a later time, or expanding it by adding further I/O modules and programming. Again, this flexibility or modularity makes this device suitable for a very vast array of control projects.

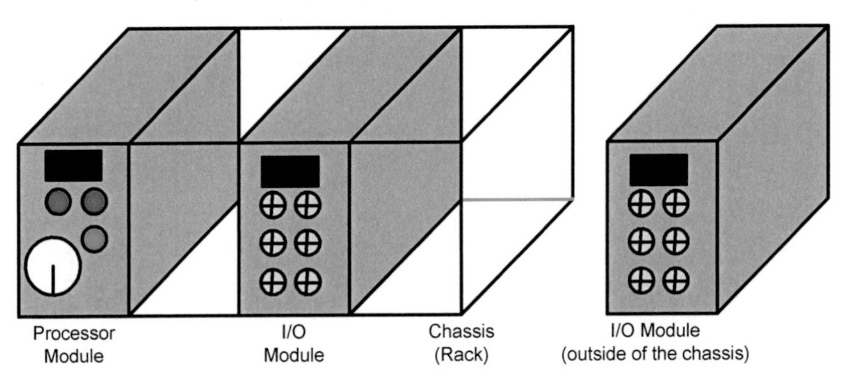

Time to bring some examples: A pressure switch, mentioned earlier will be a digital input to the PLC. A pushbutton in an Industrial panel is also a digital input. In the hypothetical case that we use a PLC to manage our house lighting, the light switch will be a digital input as well. Visualize the digital input as the one carrying out the condition in our action/condition pair. We will pause or discussion for a moment, in order to bring another concept: most industrial electrical motors are controlled by a **magnetic starter**, a special type of relay. Here the coil operates (is energized) at a rather low voltage level (usually 120VAC –Volts AC-) and the contacts are designed to allow the passage or current at a higher voltage level (usually 480VAC), with three poles, to allow the connection of the three lines pertaining to a three-phase electric power system used to run the motor. To summarize things over, if you energize the starter coil, the motor is in turn energized, and conversely, if you de-energize the starter coil, the motor is de-energized in turn. We happen to go away from the discussion and introduce this concept because the PLC connection to the starter coil is one of the most common digital outputs used in Industrial environments. To continue in our line of I/O examples, a pilot light in a panel is also a digital output. Going back to our hypothetical PLC controlling the house lighting system, the lights or lamps are digital outputs from our PLC. Visualize the digital output as the one carrying out the action in our action/condition pair.

Most process measurements in the industrial environment are performed by different measuring **probes** (for example, but not limited to: temperature, pressure, pH, other analysis like conductivity, gas Oxygen content, electric current amps or weight). These probes are usually connected (or sometimes built in together in a single unit) to a **transmitter**, an electronic device that manages the electronic signal produced by the probe and transforms the signal into another signal that is **proportional** to the measurement, so that if the measurement increases or decreases in magnitude, the signal (current or voltage) generated by the transmitter increases or decreases in the same fashion. Let's say we have a temperature transmitter calibrated for a range of 0°C to 100°C, with and output signal of 4 to 20 milliamps DC (mADC). When the probe is measuring 0°C, the transmitter generates an output signal of 4mADC. If the temperature happens to rise to 100°C, the transmitter in turn will generate an output signal of 20mADC. If the measured temperature is half-scale or at 50°C, the transmitter will generate a signal halfway in the 4 to 20mADC range, or 12mADC. The signal will vary in a linear proportion with the measured temperature. This output signal of the transmitter (usually this mentioned range of 4 – 20mADC;

another usual one is 1 – 5VDC, among others) is connected to an analog input on the PLC system.

Very usual in industrial process control are **control valves**, sets of equipment composed of a valve and a **positioner**, a device capable of giving a controlled mechanical movement and position based on an analog signal connected to it. For example a valve/positioner set accepting a 4 – 20 mADC signal, will produce movement such that the valve will be fully closed (or at 0%) when 4mADC are applied, and fully open (or at 100%) when 20mADC are applied. Halfway through the range, the valve will be exactly half open (or at 50%) when half the signal, or 12mA are applied. The valve position will vary in a linear proportion with the applied signal. This signal connected to the positioner is an example of an analog output from our PLC system.

To conclude with our examples, there is an interesting situation very often found in industrial environments, when we need to vary the speed of a motor (a starter will just drive to its full rated speed and will not allow speed variations), we use an electronic device called a **Variable Speed Drive (VSD)**. Similar to the starter, we energize the driven motor by energizing a Digital Output connected to the VSD. When the motor is energized, the VSD gives us a **feedback signal** (closed or energized when the motor is running; open or de-energized when the motor is stopped), which is a Digital Input to our PLC system. We also need to command the speed of the motor. For this purpose, we use an analog signal, provided by one of the Analog Outputs of our PLC system. The VSD, in turn, gives us another feedback signal proportional to the actual motor speed, which we usually connect to an Analog Inputs of our system. So, here we have a device that uses all four basic instances of our I/O. The point here, besides the practical considerations, is that the combinations of Inputs and Outputs for real-life cases are limitless.

Programming Software

Let's go back to our wizard computer itself. In order for it to substitute the relay arrangement and for us to gain access to all the advantages discussed, we need to design some code, a program, usually in the form of a ladder diagram, for this wizard computer to execute. We design the ladder code for our intended application, and now go ahead to discuss our means to transmit this ladder diagram we need for our application to the **PLC**

Processor Memory. We do need a way to interact with our friend, in order to write program code and transmit it to the PLC memory, edit it if needed, and troubleshoot in both the cases that our program code may be incorrect or that the controlled process has some problem. We also need to somehow visualize the actual data in the processor memory, and sometimes even manually modify it. For these purposes, we use specially designed software, usually referred to as **Programming Software**, and also usually supplied by the PLC vendor. We run this software in a computer that is somehow communicating with the PLC processor. For this, we use one or more of many available communication network types that will later be discussed in detail. Having this capability of communicating with our processor, we then have two modes of programming our code: First, let's say that we are not actually using the communication link between the computer and the PLC, and we are rather developing some code concept, without interacting with the PLC memory: this is **Offline** mode. Conversely, we can also actively make use of the communications link and interact directly with the processor memory: this is **Online** mode. When we have the case of a new application, we often develop our application Offline before the installation of the processor, and then transmit the complete application to the processor memory: this is called **downloading** to the PLC memory. When we perform a download, we erase all the PLC memory, and substitute what used to reside in that memory with the application we are downloading. Some processors (not that many) offer the capability of a **partial download**, where we decide which portion of the memory is erased and substituted. On the other hand, when we have an existing application that we need to modify or troubleshoot, we go Online with the processor, view and interact with the processor memory contents, and, as the changes to an existing application are often relatively less complex, we can modify the program code online as well. Which kinds of operations are allowed to be performed Online depends on the type of processor. Usually, major memory structure modifications are not allowed, while modifications to the ladder code are. As a rule of thumb, the smaller (sometimes also the older) the processor, the lesser Online modifications are allowed, up to the point of some very small (or some very old) processors that do not allow Online modifications, and each time we need to modify our code, a download is needed.

There is also the possibility that we need to work on an existing application, which has been already downloaded to the PLC Processor memory (sometimes even to copy the application a deploy it to other PLC Processors). In this case, we will need for our connected computer to read the PLC Processor memory and transfer the application to the connected computer. This operation is called an **Upload**.

Sometimes this Download/Upload nomenclature can become somewhat confusing. We can suggest the memory aid to imagine that the connected computer is always atop of the PLC Processor. Then, a Download will make data flow in the downwards direction (from the connected computer to the PLC Processor memory), and conversely an Upload will make data flow in the upwards direction (from the PLC Processor memory to the connected computer).

Operation of the PLC Processor

In order to get prepared to develop PLC code and start solving real-life application issues, we need first to have some understanding of how this wizard computer operates. As with most computers, it all comes down to data: our friend receives data from the outside world in the form of Inputs, sends data to the same outside world in the form of Outputs, and also manages some data internally in order to perform the required operations. This data resides in the PLC processor memory. The specific arrangement of this memory is the subject of our next topic. The PLC receives its initial application via download, or its modifications also via download or Online modifications; then we need to decide if the processor will actually execute the application (and modify or operate the connected Outputs, with the personnel safety concern that could arise from it), or will remain "dormant" (with a loaded application, but not executing it). To execute the application, we place the PLC in **Run Mode**. In order for the PLC to remain "dormant", we place it in **Program Mode**. Some larger (or newer) processors have also available a special troubleshooting mode, where the application runs, but the Outputs are not operated as a result. This special troubleshooting mode is **Test Mode**. The extent of Online modifications allowed by the PLC processor also depends on its mode, as some processors allow memory and other major modifications Online, but only while on Program Mode. Most PLC processors have means to switch from one Mode to the other as switches (often key switches, to prevent unauthorized personnel to have access) in the processor physical module. In order to allow these Mode change operations from the Programming Software, usually this key switch has a "Remote" setting, meaning that the Mode control is being performed remotely, from the Programming Software running in a connected computer. The Programming Software shows us the Mode of the processor as either Remote or Local, and either Program, Run, or Test. In order to summarize this discussion, we present the following Mode Table:

Processor Mode	Remote	Local
Run	Software shows **Remote Run**: • Code is executing. • Outputs are operated. • Minor Online modifications are allowed, depending on processor type.	Software shows **Run** (local is not shown): • Code is executing. • Outputs are operated. • There is no control (only monitoring) from a connected computer.
Program	Software shows **Remote Program**: • Code is not executing. • Outputs are not operated. • Some major Online modifications are allowed, depending on processor type.	Software shows **Program**: • Code is not executing. • Outputs are not operated. • There is no control (only monitoring) from a connected computer.
Test	Software shows **Test**: • Code is executing. • Outputs are not operated. • Usually no modifications are performed while on this mode.	This mode is not available in Local mode.

We discussed the case where everything performs as expected, and we develop a perfect program code, download it to our processor, and the code applies perfectly to our situation. But we do know that in real life this is not often the case, problems happen, and mishaps occur. When the processor tries to execute the downloaded application and finds a major problem, or when the case arises that the data managed does not fit the application, the processor tries to recover by itself from the problem. If this recovery is not possible, the processor stops executing, a **Processor Fault** is generated, and the processor goes to **Faulted Mode**. The way to recover from this situation is to go Online with the processor; the programming software will guide us to the description of the Fault, and usually where in the code the Fault occurred. With this information, and with a review of the code, we should then evaluate the cause of the Fault. When the cause is found, then we clear the Fault. The processor goes to Program mode. Here we usually make code modifications to take care of the situation and to prevent further occurrences of the same, and then switch the processor back to Run mode to verify if the mishap is taken care of. In the Special Topics section, we will discuss a method that some processors allow to try to programmatically (code that will run automatically) recover from a processor Fault.

Going back to the programming software, it will provide us with an editor where we can enter code either Online or Offline (to be downloaded). On writing the code, we are just commanding the PLC processor to perform operations on the data it manages. We need to decide and design what the processor will do with the data, and in which order, so that we obtain a result that fits our situation. This means that we obtain the correct actions happening as a result of the conditions encountered. These data operations are performed in the form of instructions that are placed in the ladder code. Each line of code is called a **rung** (as the rungs of an actual ladder). We will discuss the instruction set in detail in an upcoming section. Besides the correct instructions, also the placement of these instructions in the rung will determine the final result, or again which actions will happen according to which conditions. If we place the condition instructions right beside each other, which means placing them in series connection (remember the analogy between the ladder diagram and the electrical connections), then we are saying that these instructions are part of a logical **AND**. This means that all of the conditions should be happening at once for the corresponding action to execute. Conversely, if we happen to place the condition instructions in parallel lines of code inside the same rung, or **branches**, then we are saying that these instructions are part of a logical **OR**. This means that any (just one or more) of the conditions should be happening for the corresponding action to execute. If we place the action instructions in **branches**, this means that all of these multiple actions will occur if the corresponding conditions are fulfilled. It is worthwhile to mention that, with very special exceptions that are beyond the scope of this text, the logic is evaluated by the processor in a left-to-right and top-to-bottom fashion in the ladder diagram. Sorry, we are not to perform any actual programming yet, as we need to discuss a couple more subjects that we need to understand before entering in any actual programming feat (think of it as if we were saving the fun part for the end, or as the "dessert").

2 PLC MEMORY

As every computer, our friend PLC needs a memory to operate, a "scratchpad" in order to perform the operations, either simple or complex, that we assign it to do in order to complete our "perform action based on a condition" tasks.

The area of PLC memory that we are to discuss (and the one that is under our practical interest, although it is not the only) is the area that we are able to modify and manage. This area of memory is divided in two segments: the first one for Program Files, where we store our ladder code (programs of ladder code, composed of rungs, and these rungs composed of instructions). We can configure only one or many Program Files, allowing us to design the structure of our PLC code. Later on, we will discuss some techniques of structured programming, for which the ability to have and address many Program Files is instrumental. Usually there is a fixed, default, "Main" Program File, which always contains code, and from there we can instruct the code to "jump" to other Program Files, which we can understand as "subroutines", or auxiliary Program Files to the Main. Usually, these Files are numbered from 2 to a limit that depends on the Processor type. File number 1 is reserved to Processor "administrative" operations. File number 2 is reserved for the Main Program File. Again, from ladder code Program File 2 we may call additional routines residing in Program Files 3 to the Processor limit.

The second memory segment is dedicated to data. In its most general sense, this Data segment is composed of words, which in turn are composed of bits. Depending on the system, a word may contain different quantity of bits. It is very common to find words composed of 16 bits, while for some large systems, words can be composed of 32 bits. This will be the actual "scratchpad" of the PLC operations. In order to perform these various and distinct operations, we need to "sub-classify" this amount of data into different Data Types. The Data memory segment is divided in Data Files. Usually, a Data File is assigned and its data is organized on a fashion that corresponds to a single Data Type. There are some kinds of these Data Types used to express the data collected directly from or to be sent to I/O modules in an understandable way. There are also other Data Types that are used to express the data (usually internal registers to auxiliate our Program tasks) also in an understandable and usable way. We go ahead to discuss the more common data types, along with their mnemonics and the basics on how to address them:

1. Input Data, I : used, as mentioned above, to express the data received from the "outside" world throughout Input modules in an understandable way. If the data comes from a Digital Input module, normally a word represents an entire Input module, where each bit sequentially represents an Input channel (a connection from a single device, for example, a pressure switch) from the module. Care should be taken when corresponding actual PLC data to physical module channels, as often, the numbering system of the PLC begins at 0 rather than 1, so the first channel is 0, the second is 1, and so on. So we must bear in mind, for example, that the fifth physical channel connected to the module corresponds to bit 4 in the Input Data word. If the Data comes from an Analog Input module, here a word represents the quantity read from each channel (again, a connection from a single device, for example, a transmitter) in the module.

Now a little bit on Data File addressing, which becomes important as this is the notation we use to refer an operand in a ladder instruction to a specific location in the Data File memory: We usually use the letter mnemonic to identify the Data Type, followed by a sequential number that identifies the File. Here, when we refer to I/O data files, there is a single previously assigned data file number, which is not changeable, and sometimes is omitted. For example, Input data file is 1. After the identifier and the File number, a separator character is used (usually a colon) to separate the File locator from the actual data locator. After the colon, another sequential number (that also usually begins counting at 0) identifies the specific word (or structure: there are more "complex" data types, which result more beneficial to organize in structures. These structures are no other that convenient groupings of words and bits; the specific structure compositions will be discussed along with the instructions that address to each Type in the Instruction Set section. There is an exception, where the simplest structure is the group of words under the same address that are used to identify the channels for an Analog Input module). If we happen to refer to a specific bit in the word or the structure, we use another separator character (now usually a forward slash). Again, for the case of Digital Input modules, a word represents an entire physical module; the sequential number for this word represents a physical position in the I/O chassis, and a bit each channel. So, for example, if we are to address the first Digital Input channel connected to a module located in the second I/O rack slot (commonly the first slot, numbered 0, is reserved as the default and only position for the processor module), we use I:01/00. Note that we don't use the number 1 to identify the Input file, as 1 is the one and only Input file and is implicit. In the same line, if we are to address the fourth channel connected to a Digital Input module located in the sixth slot of the chassis, we use I:05/03.

For the case of Analog Input modules, again there is a simple structure composed of one word for each channel, also numbered beginning from 0. So, if we are to address the first channel of an Analog Input module located in the second slot in the I/O chassis, we use I:01.00 to identify the entire word. For a second example, to address the third channel for an Analog Input module located in the fourth I/O chassis slot, we use I:03.02.

2. Output, O : in order to use an analogy with the already discussed Input Data Type, the Output Data Type is used to express the data to be sent to the "outside" world throughout Output modules in an understandable way. If the data goes to a Digital Output module,

normally a word represents an entire Output module, where each bit sequentially represents an Output channel from the module. If the Data goes to an Analog Output module, here a word represents the quantity to be written to each channel in the module. Following the fashion discussed for the Input modules, if we are to address the first Digital Output channel connected to a module located in the second I/O rack slot, we use O:01/00. If we are to address the fifth channel connected to a Digital Output module located in the seventh slot of the chassis, we use O:06/04.

For the case of Analog Output modules, again there is a simple structure composed of one word for each channel. If we are to address the first channel of an Analog Output module located in the second slot in the I/O chassis, we use O:01.00 to identify the entire word. To address the second channel for an Analog Output module located in the fifth I/O chassis slot, we use O:04.01.

3. System, S : This is a special structured, rather unmodifiable (can be read by logic, but mostly not written to) Data file, bearing but again not being written, number 2. This Data file, in general terms, contains processor "administrative" data. This data includes the processor status, **Real-Time Clock (RTC)** data when available, program scan times, faults, Run or Program as well as Remote or Local status, special mathematical status indicators, a bit that indicates that the first scan after the processor being put in Run mode is occurring, and many more of the sort. In order to use System data as ladder instruction operands, every processor family has its own details, so we would say that is better to refer to either the processor manual in each case, or to the Data File view in the programming software. Here this data is organized by classification, and a specific piece of information is somewhat easy to find. As an example, we present here the RTC data structure for a particular processor:
    - S:37 Year (4 digits)
    - S:38 Month
    - S:39 Day
    - S:40 Hour
    - S:41 Minute
    - S:42 Seconds
    - S:53 special data word for day of the week, where 0 stands for Sunday, 1 for Monday, and so on up to 6 for Saturday

4. Binary or Bit, B : This is the first Data File where the organization of the data is completely accessible to the programmer, because it does not depend on I/O or processor configuration. The B file consists of a grouping of bits with the purpose of being used as internal or auxiliary program registers, and, of course, addressable on a bit-by-bit basis. Nevertheless, these bits are organized in 16-bit words, and then we come up to have two modes of addressing: first, a direct bit addressing mode, where the bits are sequentially numbered from 0. Here, we don't use any word numbering or delimiter; we go directly from File descriptor to bit delimiter and bit sequential number. For example, the tenth bit in the array for Bit File 3 (the default B file) will be B3/09 (remember that the first bit is numbered 0). The second addressing mode uses the word number, again being numbered from 0. So, this very same tenth bit in the array will be B3:00/09, and the programming software accepts either addressing mode without problems. If we go further ahead into the Bit File array, sometimes we can find the direct bit addressing mode somewhat inconvenient, as some math is needed. For example, the hundredth bit in the array for the same File B3 would be B3/99. It becomes easier if we happen to know that this very same bit is the fourth bit of the seventh word: then it would be B3:06/03. We can convert from one mode to the other by dividing the bit number by 16, obtaining the whole number and then noting the reminder as the bit number. An example will make it simpler: If we divide 99 by 16, we obtain a result of 6 and a reminder of 3, hence B3:06/03. Conversely, we obtain the bit for direct addressing by multiplying the word number by 16 and then add the bit number. 16 Times 6 is 96, and then we add 3 and obtain 99, going back to our original B3/99. Again, File number 3 is the default File for Binary or Bit Data. These default files cannot be changed from one Data Type to other, so always, file 3 will be type B.

5. Timer, T : Our first programmer-configurable structure file. Each member of this Data File (again, numbered from 0) holds the necessary data to support an instance of a timer in the ladder logic. The specific structure of the T Data File will be discussed in the Instruction Set, where the timer instructions will be discussed, as it will make more sense when the members of the Data structure are presented. File number 4 is the default File for Timer Data.

6. Counter, C : Quite similar to the Timer Data File, this is another structure file. Each member holds the necessary data to support an

instance of a counter in the ladder logic. Again, the specific structure of the C Data File will be discussed in the Instruction Set, where the counter instructions will be discussed. File number 5 is the default File for Counter Data.

7. Control, R : This is a special structure file needed to support certain advanced instructions. We think that these advanced instructions are beyond the scope of our introductory effort, so it is mentioned here for informational purposes only. File number 6 is the default File for Control Data.

8. Integer, N : This is the simpler structure to store and express analog (or numeric) quantities. Each N address is a group of 16-bit words, each used to store one number. Out of this 16 bit structure, one bit is used for sign, and fifteen bits for the number itself. This leaves us a numeric range of from -32,768 to 32,767 to express quantities in Integer Data Type. Care must be taken when using Integer Data Type for mathematical calculations where the result could possibly exceed these boundaries, as problems with Processor faults could arise. File number 7 is the default File for Integer Data.

9. Floating Point, F : This File stores data in floating point format, which can overcome the limitation that sometimes we could find with Integer-only figures. Each quantity is stored in a structure consisting of two 16-bit words, giving us a range of positive or negative $\sim 1.175 \times 10^{-38}$ to $\sim 3.402 \times 10^{38}$. File number 8 is the default File for Floating Point Data.

10. Additional Data Files: Data Files from File 9 on, up to different limits depending on the Processor type, sometimes up to 999, are configurable by the programmer. There is the possibility to choose between Data Types, B, T, C, R, N, F for these Files (Input, Output and System Files must be unique). Also, the Data File length (number of structures in a File) is configurable by the programmer, up to a limit that also varies by Processor.

Enough for the time being with Processor data files, but not yet (oh my!) into programming. Still a couple more discussions to go before the fun part is in. Let's tackle them!

3 "ANATOMY" OF A PLC SCAN

Besides the discussed physical characteristics of a PLC that gives the system its modular structure and therefore its flexibility and suitability for multiple applications, another very singular characteristic of the PLC is its execution speed. We can build and configure very extended and complex ladder code programs, and they will be scanned completely by the processor in the order of hundreds, and even a thousand of times a second. The scan time depends on the processor type and size, and, of course, in the size and complexity of the program code.

It is worthy to make a parenthesis here and discuss yet another very singular characteristic of PLC systems. Very often, PLC processors are used to execute control of mission-critical applications where personnel and equipment safety are a must to consider. Very often as well, emergency devices (like emergency stop switches) are implemented by means of a PLC input. There are also situations where physical **interlocks** are implemented in the PLC code. An interlock is the action of disabling a device based on some condition (a perfect example of our action/condition pair). For example, if we have a vessel where process steam and process compressed air are served, typically we don't want for both process fluids to be injected to the vessel simultaneously for safety reasons. Then we put in some code that, when the steam valve is detected open, will disable the open command for the air valve, and vice-versa. Another example is the situation where we have a series of conveyors one discharging into the following one. If a downstream conveyor stops and the upstream ones do not, material will be discharged with no possibility of being cleared. Here then we place code that will disable the operation of one conveyor if the one where it discharges is not operating (sometimes this is called **cascading devices**). Being used so commonly for these critical applications, we do need for the program scan to actually "happen" or execute. Now imagine that we make some mistake in our program, where we place, for example, "infinite loops", or sections of code that keep repeating themselves without allowing the execution of further code. To avoid this "infinite loop" from prohibiting our safety code from executing, the processor has a configurable "watchdog timer". This is an allotted amount of time for a full program scan. The processor is keeping track of the time it takes for it to execute every scan. If a scan time exceeds this allotted time, the processor declares itself in fault, and usually inhibits further program processing.

Now let's go back to our scan discussion: we will try to summarize the operation as follows:

- First, we need information from the "outside world" in order to execute our program. So, the first activity is to read the connected Inputs.

- Second, now comes the actual execution of the ladder code. We will make another note here, to identify another special characteristic of some processors: we can configure a special routine called a **Selectable Time Interrupt** (STI). This ladder routine will execute on a time schedule basis (for example, every second), and will interrupt whichever part of the program scan that is executing. After the STI execution completes, and then the "normal" program execution continues. A clarification would also be in order here to mention that this "normal" program interruption does not exist in some large systems that are capable of **multitasking**; this is, running more than one process or program at the same time.

- Third, we use information from the program execution to write to the connected Outputs. This will close the loop of conditions/actions from and to the "outside world".

- Now it's time to perform processor "administrative" activities, like clearing internal memory buffers (usually not accessible to the program), determining the scan time and compare it to the watchdog time setting, and basically getting ready for the next scan. Then, everything is started all over again.

4 INSTRUCTION SET

Our wizard friend being introduced and discussed, the Basic instruction set is to be discussed, trying to remain as neutral as possible in terms of system manufacturer. These instructions become the building blocks for our program applications. These basic instructions are to be found in most systems, with minor variations in naming and symbols or mnemonics (abbreviations), and mostly with the same functionality. As often as possible, ladder examples will be introduced in order to clarify the usage and operation of the instructions, as well as to present its practical applications.

A. Basic Bit Instructions:
   1. Normally open contact, XIC, -| |- : This is the basic building block or our ladder logic library. It represents the basic "switch" function: if a switch is on, then energize the connected circuit. A normally open contact will evaluate to True (will allow the logic to its right hand in the ladder diagram rung to become true) if the condition that it represents (can be a connected input or a storage bit) is in turn true, and will evaluate to False (allow the logic to its right to become false) if the condition it represents is false. The normally open contact is an input instruction (represents the condition in the action/condition pair), and thus is written in the left section of the ladder rung. Usually, the input instructions deal with memory addresses, or directly with Inputs connected to the PLC.
   2. Normally closed contact, XIO, -|/|- : This is the counterpart and opposite of the normally open contact. In terms of the "switch" function, if a switch is off, then energize the connected circuit. As opposed to its counterpart, a normally closed contact will evaluate to

True if the condition that it represents is false, and will evaluate to False if the condition it represents is true. The normally closed contact is an input instruction.

3. Coil energize, OTE, -( )- : This is the basic logic result instruction in our library. A coil energize will become true (can turn on a connected output) if the logic to its left hand in the ladder diagram rung is true. The instruction will become false (can turn off a connected output) if the logic to its left hand is false. The coil energize is an output instruction (represents the action in the action/condition pair), and thus is written in the right section of the ladder rung. Usually, the output instructions deal with memory addresses or directly with Outputs connected to the PLC.

Ladder Code Example 1:

Consider the situation to start or stop a motor with one start and one stop pushbuttons. We need for the motor to stay energized when the operator releases the start pushbutton, and to de-energize when the operator pushes the stop pushbutton.

Solution:

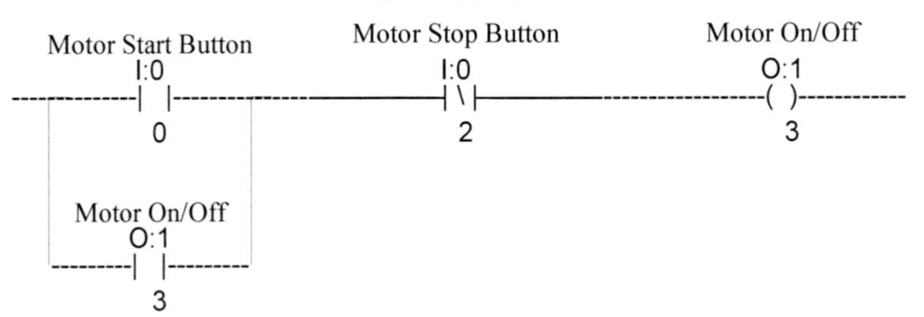

Code Explanation:

When the Start pushbutton, connected to I:0/0, is actuated and the Stop pushbutton, connected to I:0/2, is not actuated (note the XIO instruction and remember that it evaluates to True if the connected Input is de-energized), the Motor On/Off Output, connected to O:1/3, is energized. Note that an XIC instruction is branched in parallel with that corresponding to the Start pushbutton, referring to the same Output as the Motor On/Off contact. This means that as soon as the Output energizes, this instruction evaluates to True (note also that we can use almost any Data Table bit memory address, in this case an Output address, as the XIC reference). This instruction evaluating to True will allow the Output to remain energized when the operator releases the Start pushbutton. This is commonly referred as "**sealing**" or **latching** a contact (for it to remain energized even after the condition that triggers it is not True anymore). When the Stop pushbutton is actuated (the XIO instruction evaluates to False), the Output is de-energized, and the "seal" is broken (the Output will not energize again when the Stop pushbutton is released, rather needs for the Start pushbutton to be actuated again).

On presenting the first ladder example, we pause to present an important point: please note that every instruction in our ladder example has a comment (in this case over its representation, there can be variations among ladder editors from different suppliers of processors) that helps us to identify it, and somehow gives us a hint about its function. Also, there is a comment atop the rung representation, summarizing the operation performed in this rung. This is the **documentation** of the ladder code, and its usage constitutes a good practice when developing ladder code. We need to visualize that if we happen to design and develop some code, it is not necessarily true that we will also be the only ones interacting with the software: monitoring, modifying, and troubleshooting problems with it. Some other person that needs to interact with the system would not necessarily have the same understanding of the code that we had when we developed it. Even we can forget some details if it happens for some time to elapse before we need to interact with our code again. In order to avoid all of these situations, it is quite healthy to perform good code documentation practices, and to try as much as we can to document every instruction and every rung of our code.

4. Output latch, OTL, -(L)- : An output latch will become true when there is a **false-to-true transition** (logic was false and becomes true) of the logic to its left, and will remain true (or energized) until a Output Unlatch instruction is executed on the same memory bit address. This is used when we want for a memory address (or a connected output, in turn) to become energized as a result of some logic condition, but to remain energized regardless of the further

state of that logic condition. The output latch is an output instruction.

5. Output Unlatch, OTU, -(U)- : The counterpart and complement of the output latch instruction. The output unlatch will become false when there is a **false-to-true transition** of the logic to its left. It is used to de-energize a memory address that has been previously latched. The output unlatch is an output instruction.
6. One-Shot, -[ONS]- : ONS is a special input instruction that, when there is a false-to-true transition of the logic to its left, allows for the logic to its right to become true for only one program scan. After this scan, the logic to the right of the ONS becomes false again, until the next false-to-true transition of the logic to its left. There are some situations where we need for an action to be performed just once when a condition happens. ONS comes very handy for use in these situations.

Ladder Code Example 2:

Consider the situation for our motor to have a cascade device (i.e. a conveyor to which our motor-driven device discharges, again with one start and one stop pushbuttons. Again, we need for the motor to stay energized when the operator releases the start pushbutton, and to de-energize when the operator pushes the stop pushbutton. We also need for our motor to operate only when the cascade device is operating.

Solution:

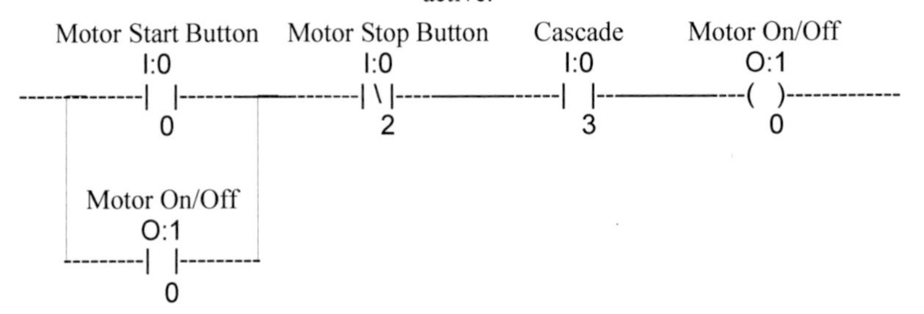

Code Explanation:

The operation of this piece of code is very similar to that of the Example 1, but with the obvious difference of the presence of the cascade contact. In order for the Output to be energized when the Start pushbutton is actuated, we need for the cascade contact to be energized (evaluating to True). Then the Output is actuated upon the Start pushbutton being actuated, the "seal" is created, but now the "seal" can be broken by actuating the Stop pushbutton **or** when the cascade contact is de-energized. This causes that when the cascade device ceases operation, our motor stops. Often, we are operating conveyor motors and the cascade device is the downstream conveyor where our conveyor motor discharges. This avoids the case of a conveyor operating, discharging into a stopped conveyor, causing either material loss or blockages, or even safety issues.

Ladder Code Example 3:

Consider the situation to fill a tank by operating a filling pump or opening a filling valve. We need for our pump (or valve) to start (or open) on the Low Level input from the Tank sensor, and to stop on the High Level input from the tank.

Solution:

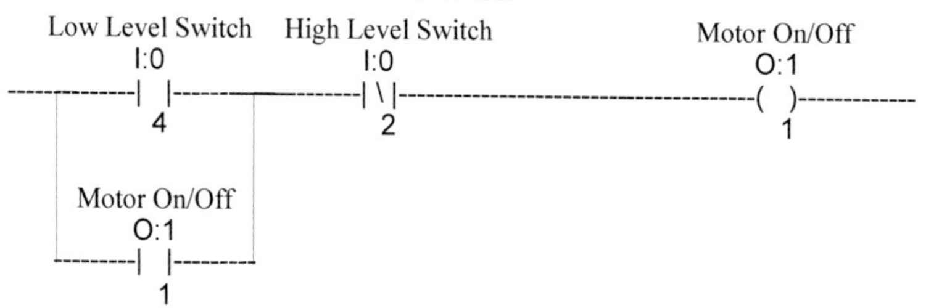

Code Explanation:

This example is better understood if we happen to compare it with that of Example 1 (note that the instructions and their arrangement are the same, giving us a hint on another characteristic of the flexibility of our friend PLC: the possibility of using similar code for different applications). When the Low Level switch is actuated (empty Tank) and the High Level switch is not actuated, our Motor starts (or filling valve opens), its Output contact is "sealed", and will remain in operation until the High Level switch is actuated (full Tank). The cycle will start again when there is an empty Tank detection.

Ladder Code Example 4:

Consider now the situation to start a reversible motor with one forward start, one reverse start and one stop pushbuttons. We need for the motor to stay energized when the operator releases either start pushbutton, and to de-energize when the operator actuates the stop pushbutton. We also need for the code to prohibit the possibility of both forward and reverse start contactors to be energized at the same time.

Solution:

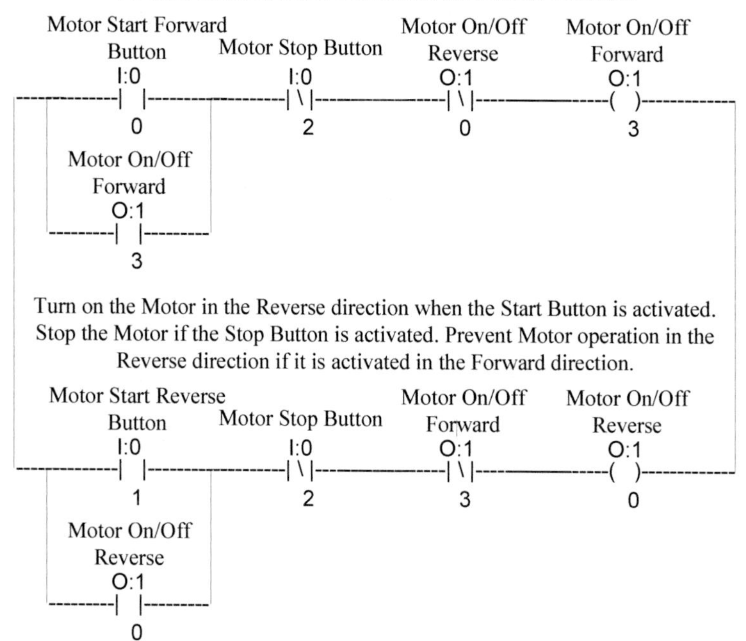

Code Explanation:

In this Example, we can refer to Example 1 to understand the Motor start/stop operation, and to Example 2 to understand how a cascade contact operates. Now think that the Start/Stop contact for the opposite direction becomes the cascade contact that prohibits the operation in each case. Note that the Output references are used in XIO instructions respectively, thus when the Outputs are energized, these instructions evaluate to False, and the motor Output contact is de-energized or prevented from operation.

B. Timer and Counter Instructions:

1. Timer On Delay, TON : Up to this moment, our instruction set discussion has been limited to instructions that act on single bit memory addresses. Later, we will also deal with instructions that act on single word memory addresses. At this moment, we need to consider a Data Structure on the PLC Data memory section. A data structure is a logical grouping of bits and words, which are used for a single (and somewhat complex) operation, and respond to a single memory address. For example, let's consider a timer, which happens to be the simpler data structure. If we were to make a timer work by ourselves, we need some kind of "scratchpad" where to keep note of the operation of our timer. We need to know at which moment the timer will end timing (and this value takes a word in the PLC memory, and is commonly called Preset value). We also need to know how much time has elapsed in order to compare this value with the preset value, and then determine if our timer needs to continue timing or not (this value also takes a word in the PLC memory, and is called the Accumulated value). A bit should tell us when to start the timer, and if the timer is not needed to keep timing anymore (this bit in the PLC memory is called the Enable bit). Another bit should stand as an indicator that our timer is actually timing (this is the Timer Timing bit). Another bit should indicate that the timer has completed timing (meaning that the Accumulated Value is now equal to the Preset Value: this bit is the Done bit). The Base entry refers to the time base of the timer, usually 1.0, 0.01, or 0.001 seconds. This means that, if we have a time base of 1.0, each timer "unit" represents a second. Hence, if we see an Accumulated value of 100, this means that 100 seconds have elapsed. The Preset value needs to be aligned with the time base. If we need for the Preset to be 10 seconds, and again the time base is 1.0, then our Preset should be 10. If the time base is 0.01 seconds and we need the same Preset of 10 seconds, the Preset value should now be 1000. If the time base is 0.001 and we need the same 10 seconds, our Preset should now be 10000. When we create a timer structure in the PLC memory, all this structure elements respond to a single memory address, as discussed before. Individual members of a structure are denominated by adding a suffix to the structure address.

For example, if we call our recently created timer T4:1, then the members of the structure are defined as follows:

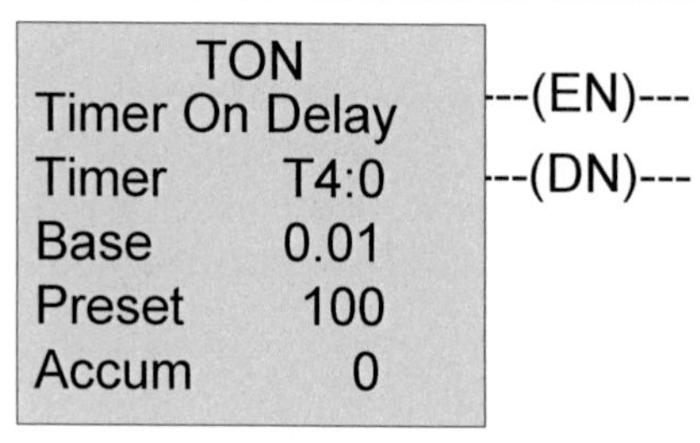

a. T4:1.PRE for the Preset Value word
b. T4:1.ACC for the Accumulated Value word
c. T4:1.EN for the Enable bit
d. T4:1.TT for the Timer Timing bit
e. T4:1.DN for the Done bit

The timer structure (as well as other data structures to be discussed later on) interacts with the "outside" world when we write additional logic to relate its structure members with actual I/O connected to the PLC. There can also be the case that we want to relate these structure members with memory addresses in the PLC Data memory.

Upon the logic to its left becomes true, The Timer On Delay instruction will delay the turning on, or energizing a bit (the timer done, or .DN bit). If the logic to the left of the instruction becomes false, the timer resets itself: this means that the timer does not time anymore, and the accumulated values goes to zero, so the next time the logic to the left of the instruction becomes true, the timer will begin again timing from zero. When the logic to the left becomes true, the .EN bit is set, the .DN bit remains reset, and the .TT bit is set. When the timer completes timing, the timer stops accumulating, both the .EN and .DN bits are set, the .TT bit is reset, and the structure will wait for the next true-to-false transition of the logic to its left to reset itself. Then, on the next false-to-true transition of the logic, it will begin operation again. The Timer On Delay is an output instruction.

2. Timer Off Delay, TOF : Conversely to the Timer On Delay instruction, The Timer Off delay instruction will delay the turning off, or de-energizing a bit (the timer done, or .DN bit). The TOF instruction also uses the same timer data structure defined above, and also resets itself when the logic to its left becomes false. When the

logic to the left becomes true, the .EN bit, the .DN bit, and the .TT bits are set. When the timer completes timing, the timer stops accumulating, the .DN, or done bit is reset, the timer stops timing, the .EN bit remain set, the .TT bit is reset, and the structure will wait for the next true to false transition of the logic to its left to reset itself. Then, on the next false-to-true transition of the logic, it will begin operation again. The Timer Off Delay is an output instruction.

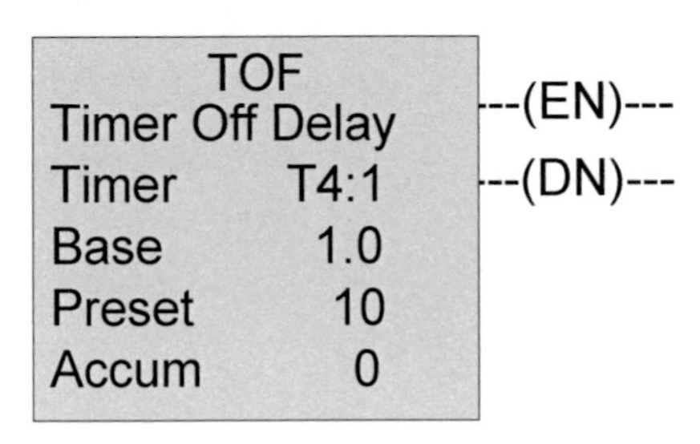

3. Retentive Timer On, RTO : This is a special variant of the TON instruction: remember that if we happen to use a TON instruction and the logic to its left becomes false, the timer data structure will reset itself. Here, if the logic to its left becomes true, the RTO instruction will begin timing the same as a TON instruction. But, if the logic to its left becomes false, the RTO will stop timing, but will not reset itself. So, when the logic to its left becomes true again, the RTO will resume timing from the last value of .ACC, or accumulated value. This instruction is used when we want to time a delay that will not necessarily be consecutive (for example, we want to measure the time the level of a Tank is above 50%, but the level rises and lowers, so the logic to the left of our RTO will alternate between true and false). When the logic to the left becomes true, the .EN bit is set, the .DN bit remains reset, and the .TT bit is set. When the timer completes timing, the timer stops accumulating, both the .EN and .DN bits are set, and the .TT bit is reset. Resetting of the timer structure is carried out by the use of a Reset instruction that will be discussed later on. The Retentive Timer On is an output instruction.

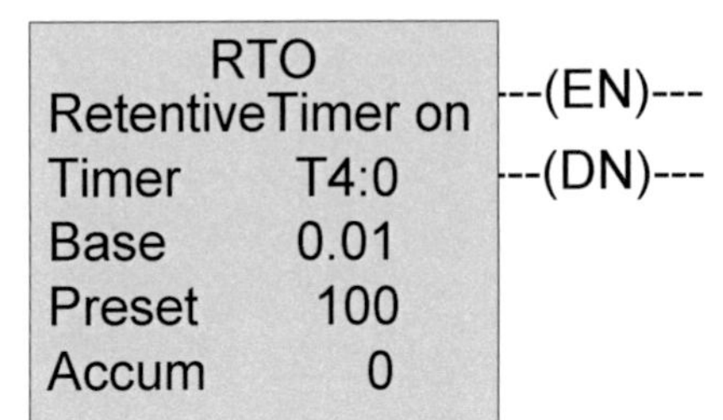

Ladder Code Example 5:

Consider a simple situation which we find often in our everyday life: a traffic light.

Solution:

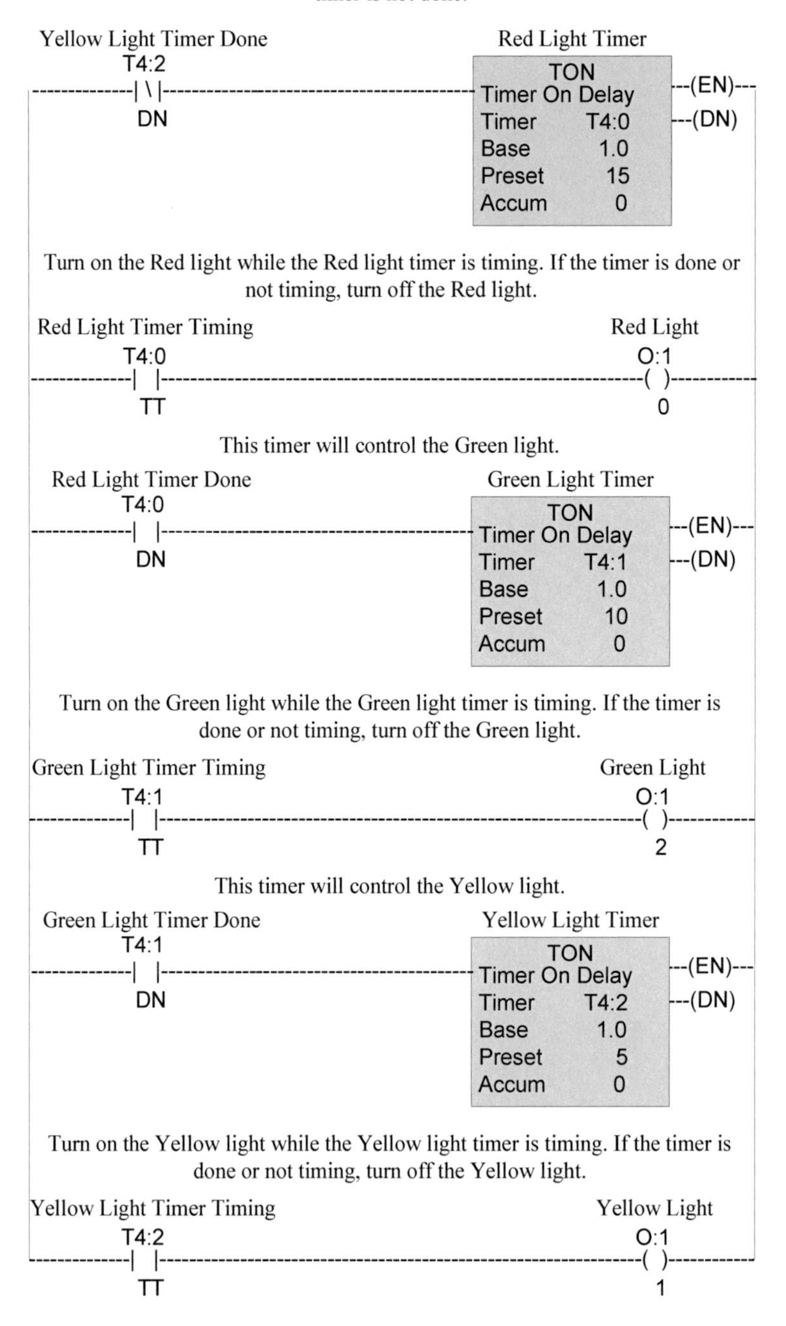
This timer will control the Red light. It will initiate the cycle if the Yellow light timer is not done.
Yellow Light Timer Done
T4:2
DN
Red Light Timer
TON
Timer On Delay
Timer T4:0
Base 1.0
Preset 15
Accum 0
(EN)
(DN)
Turn on the Red light while the Red light timer is timing. If the timer is done or not timing, turn off the Red light.
Red Light Timer Timing
T4:0
TT
Red Light
O:1
0
This timer will control the Green light.
Red Light Timer Done
T4:0
DN
Green Light Timer
TON
Timer On Delay
Timer T4:1
Base 1.0
Preset 10
Accum 0
(EN)
(DN)
Turn on the Green light while the Green light timer is timing. If the timer is done or not timing, turn off the Green light.
Green Light Timer Timing
T4:1
TT
Green Light
O:1
2
This timer will control the Yellow light.
Green Light Timer Done
T4:1
DN
Yellow Light Timer
TON
Timer On Delay
Timer T4:2
Base 1.0
Preset 5
Accum 0
(EN)
(DN)
Turn on the Yellow light while the Yellow light timer is timing. If the timer is done or not timing, turn off the Yellow light.
Yellow Light Timer Timing
T4:2
TT
Yellow Light
O:1
1

Code Explanation:

When the Yellow Light timer is not done (either at the beginning of the cycle or after the Yellow Light is turned off), the Red Light timer begins timing, and the Red Light is energized while this timer is timing (because of the use of the .TT reference). When the Red Light timer is done, the Red Light de-energizes (because the timer is not timing anymore), the Green Light timer begins timing, and the Green Light is energized while this timer is timing as well. Note that the Red Light timer remains done, because the logic to its left is still true (the Yellow Light timer is not done). The process is repeated: when the Green Light timer is done, the Green Light de-energizes (because the timer is not timing anymore), and the Yellow Light timer begins timing, and the Yellow Light is in turn energized while this timer is timing. When the Yellow Light timer is done, there is a kind of a domino-effect: the Red Light timer resets, its Done bit is reset as well, so that the Green Light timer is reset, again its Done bit is reset, so that the Yellow Light timer is reset. The Yellow Light timer done bit is reset, thus initiating the operation of the Red Light timer, and the cycle all over again.

Ladder Code Example 6:

Consider another from everyday life: some pickup trucks have only one rear indicating light on each side, acting both as stop light and turn light. But, what happens if we activate the turn lights while depressing the brake? We need for the turn light to interrupt the command from the brake switch to turn on the lights in a steady fashion, but only for the side where the turn light will be activated. Of course, the brake light is steady, while the turn light is flashing.

Solution:

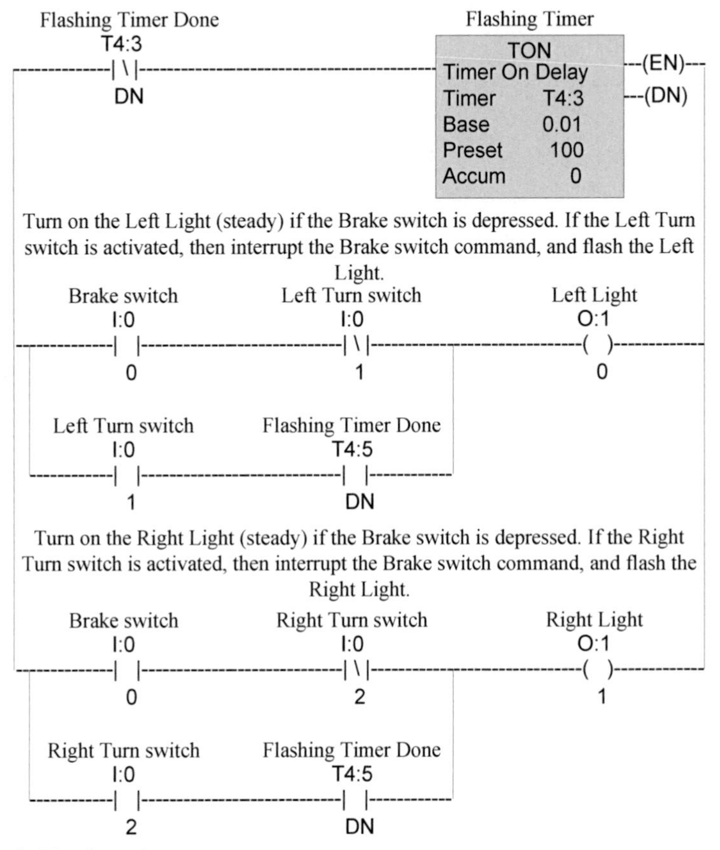

Code Explanation:

The timer on the first rung is a self-resetting timer (the Done bit XIO interrupts the timer, resetting it, thus de-energizing the Done bit itself, so the timer will start all over again) that will control the flashing of the lights (to flash a light, just turn it on when the timer is done; to vary the flashing frequency, just vary in turn the timer preset value). The second and third rungs have the same structure, one for the left light and the other for the right light. When the brake pedal is depressed (I:0/0) and the respective turn light is not active (I:0/1 or I:0/2), then turn on the correspondent light. If the brake switch is active and the turn light is active as well, interrupt the branch where the brake switch commands, and only turn on the light when the timer is done (thus flashing it).

4. Count Up, CTU : In order to start discussing counter instructions, let's make a parallel of our timer discussion: again, for example, if we were to make the counter work by ourselves, we need some kind of "scratchpad" where to keep note of the operation of our counter. We need to know at which moment the counter will end counting (and this value takes a word in the PLC memory, and is commonly called Preset value). We also need to know how many counts have happened in order to compare this value with the preset value, and then determine if our counter has reached its "target" or not (this value also takes a word in the PLC memory, and is called the Accumulated value). A bit should tell us when to count, and (this bit in the PLC memory is called the Count Up bit). Another bit should indicate that the counter has reached its "target" (meaning that the Accumulated Value is now equal to the Preset Value: this bit is the Done bit). Now we need to introduce the second data structure in our collection: the Counter data structure. It has some similarity with the timer data structure discussed above. We have, for example, that if we call our counter structure C5:1, then the members of the structure are defined as follows:

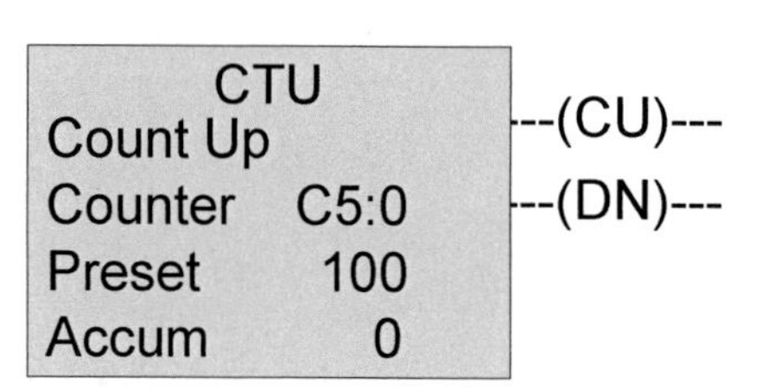

   a. C5:1.PRE for the Preset Value word
   b. C5:1.ACC for the Accumulated Value word
   c. C5:1.EN for the Enable bit
   d. C5:1.CU for the Count Up bit
   e. C5:1.DN for the Done bit

   Upon the logic to its left becomes true, The Count Up instruction increases the accumulated value by one every time the processor scans it. When the logic to the left becomes true, the .CU bit is set, and the .DN bit remains reset. When the counter reaches its "target" value, the .DN bit is set, but the CTU instruction will keep increasing the Accumulated value on every scan with the logic to its left being true. The counter data structure is reset by the use of a Reset instruction that will be discussed later on. The Count Up is an output instruction.

5. Count Down, CTD : Conversely to the Count Up instruction, the Count Down instruction decreases the accumulated value by one every time the processor scans it. When the logic to the left becomes true, the .CD bit is set, and the .DN bit remains reset. When the counter reaches its "target" value, the .DN bit is set, but the CTD instruction will keep decreasing the Accumulated value on every scan with the logic to its left being true. The counter data structure is reset by the use of a Reset instruction that will be discussed later on. The Count Down is an output instruction.

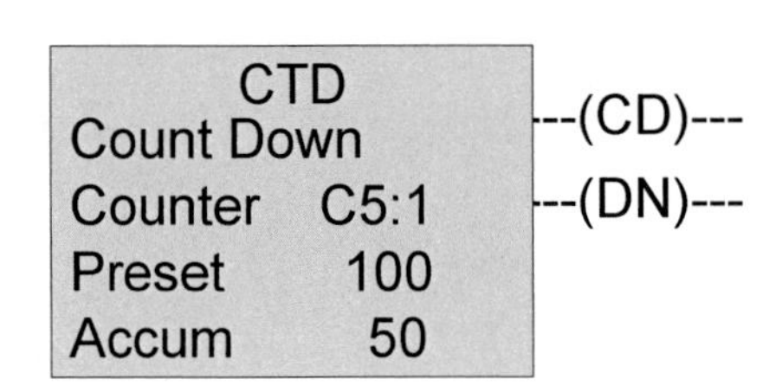

6. Reset, -(RES)- : When the logic to the left of this instruction becomes true, it actuates by resetting the either timer or counter structure to which it refers. In the case of a TON timer, upon the RES instruction acting on the timer data structure, the .EN, .DN, .TT bits are reset, the timer stops timing (until the logic again sets the .EN bit and the timer starts timing again), and the Accumulated Value goes to zero. The preset value is not modified by the RES instruction neither for a timer nor for a counter data structure. In the case of a counter, the .CU or .CD and the .DN bits are reset, the counter stops counting (until the logic again sets the .CU or .CD bit and the counter counts again), and the Accumulated Value goes to zero. It is not recommended to use a RES instruction to reset the timer referenced by a TOF instruction, as unpredictable processor behavior may occur. The Reset Instruction is an output instruction.

```
      T4:0
-----(RES)-----
```

Ladder Code Example 7:

Now consider the same situation seen before to fill a tank by operating a filling pump, but now we have two pumps to be operated and the requisite to operate one pump in one level cycle (the cycle being level being low, filling, and the level being high), and then operate the other one in the next level cycle (this arrangement is often found and called an alternator circuit). Again, we need for our pumps to start on the Low Level input from the Tank sensor, and to stop on the High Level input from the tank.

Solution:

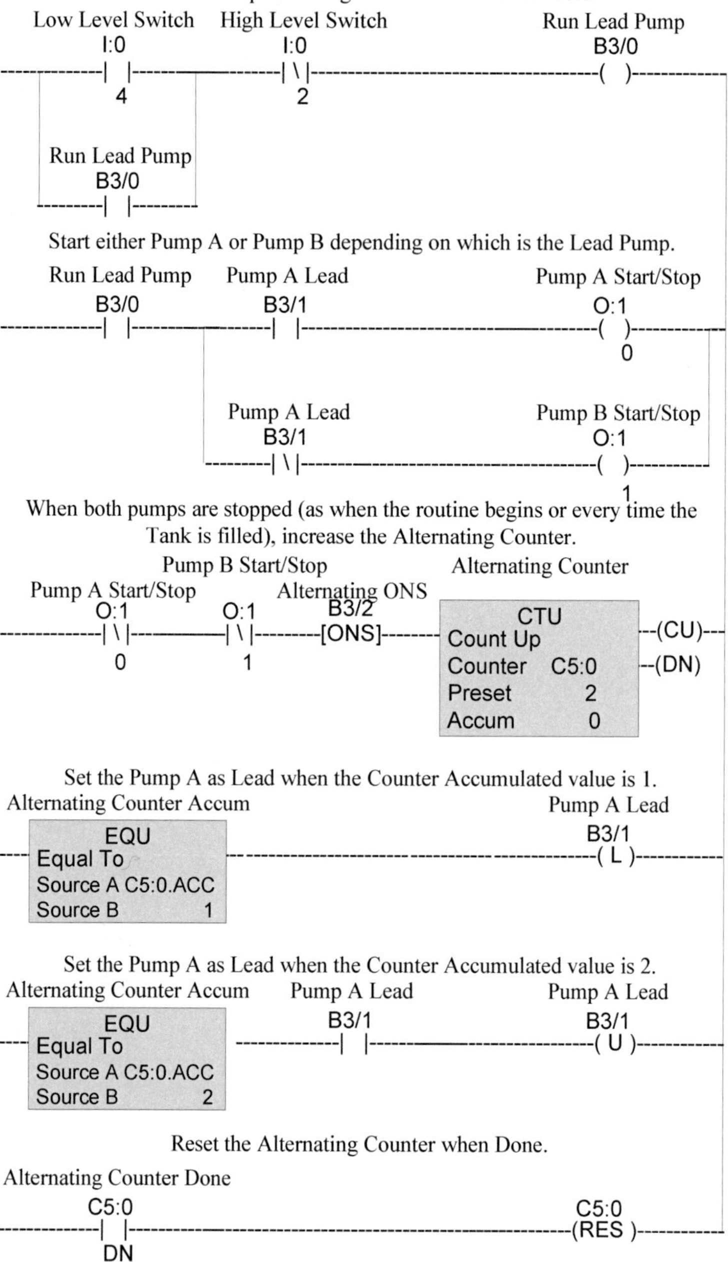
Command to run the Lead Pump when the Low Level Switch is activated. Stop the Lead Pump if the High Level Switch is activated.
Low Level Switch
I:0
4
High Level Switch
I:0
2
Run Lead Pump
B3/0
Run Lead Pump
B3/0
Start either Pump A or Pump B depending on which is the Lead Pump.
Run Lead Pump
B3/0
Pump A Lead
B3/1
Pump A Start/Stop
O:1
0
Pump A Lead
B3/1
Pump B Start/Stop
O:1
1
When both pumps are stopped (as when the routine begins or every time the Tank is filled), increase the Alternating Counter.
Pump A Start/Stop
O:1
0
Pump B Start/Stop
O:1
1
Alternating ONS
B3/2
[ONS]
Alternating Counter
CTU
Count Up
Counter C5:0
Preset 2
Accum 0
(CU)
(DN)
Set the Pump A as Lead when the Counter Accumulated value is 1.
Alternating Counter Accum
EQU
Equal To
Source A C5:0.ACC
Source B 1
Pump A Lead
B3/1
(L)
Set the Pump A as Lead when the Counter Accumulated value is 2.
Alternating Counter Accum
EQU
Equal To
Source A C5:0.ACC
Source B 2
Pump A Lead
B3/1
Pump A Lead
B3/1
(U)
Reset the Alternating Counter when Done.
Alternating Counter Done
C5:0
DN
C5:0
(RES)

Code Explanation:

The operation to energize the Run Lead Pump buffer address, B3/0, is explained in Example 3. Then, when we have this buffer address energized, we operate either Pump A or Pump B depending if the Pump A Lead buffer address, B3/1, is energized or not (energized, operate Pump A; not energized, operate Pump B). On the first time that there is no pump operating (end of a filling cycle – tank full), operate the Counter in register C5:0 (note that the ONS instruction will prevent the Counter to operate in every subsequent processor scan; this brings up to a rather important point to consider: we should always bear in mind the large processing speed of our friend PLC, so in most cases our code will be true during many –say thousands- of scans, and we should take programming precautions accordingly). When the Counter accumulated value is 1, latch the Pump A Lead buffer address. When the Counter accumulated value is 2, and the Pump A Lead buffer address is energized, then unlatch the buffer address (so that Pump B becomes Lead). When the Counter is Done (its accumulated value is 2, equal to its preset value), reset the Counter and begin the cycle again.

C. Comparison Instructions:

Now we will try to use the mathematical capabilities of our friend PLC computer (the same ones we mentioned earlier that will be difficult to implement in relay logic), and connect some logic to numeric quantities, either stored in word registers in the PLC data memory area, or coming directly from Analog Inputs.

1. Equal To, EQU : When the quantity in Source A (Source A) is equal to the quantity in Source B (Source B), this instruction evaluates to True. The instructions sounds and actually is plain simple, but care should be taken in the data types for both Sources, as we can found some trouble when we use different data types for them. For example, we could end up not having the same number (and then the EQU instruction evaluating to False), if we compare an Integer register where we have stored the

EQU	
Equal To	
Source A	N7:0
Source B	N7:5

result of the division 3/2, with another Floating Point register with the value 1.5 . We should try whenever possible to compare sources with the same data types. EQU is an input instruction.

2. Not Equal To, NEQ : This is the opposite for the EQU instruction, evaluating to True when Source A is not equal to Source B. Again, care should be taken in the data types of the two sources so we don't get an unexpected logic result. NEQ is an input instruction.

NEQ	
Not Equal	
Source A	N7:4
Source B	N7:6

3. Greater Than, GRT : When Source A is greater than Source B, the GRT instruction evaluates to True. If Source A is exactly equal or less than to Source B, this instruction evaluates to False. GRT is an input instruction.

GRT	
Greater Than	
Source A	N7:2
Source B	N7:8

4. Less Than, LES : When Source A is less than Source B, the LES instruction evaluates to True. If Source A is exactly equal to or greater than Source B, this instruction evaluates to False. LES is an input instruction.

LES	
Less Than	
Source A	N7:2
Source B	N7:6

5. Greater Than or Equal To, GEQ : This instruction works the same as the GRT instruction, with the difference that, when Source A is exactly equal to Source B, the instruction evaluates True. GEQ is an input instruction.

GEQ	
Greater Than or Equal	
Source A	N7:2
Source B	N7:5

6. Less Than or Equal To, LEQ : This instruction works the same as the LES instruction, with the difference that, when Source A is exactly equal to Source B, the instruction evaluates True. LEQ is an input instruction.

LEQ	
Less Than or Equal	
Source A	N7:2
Source B	N7:5

Ladder Code Example 8:

Let's revisit our traffic light to see if it is possible to further simplify the code solution already presented:

Solution:

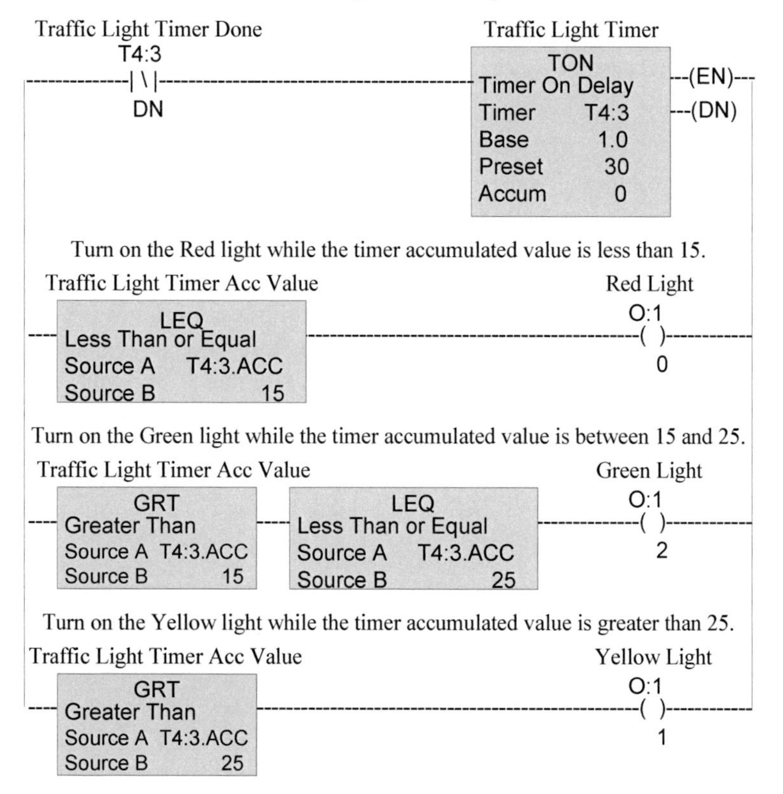

Code Explanation:

Now we simplify the solution discussed before. Again a self-resetting timer with a preset value equal to the complete cycle time (30 seconds in this case). Then turn on the lights depending on the accumulated value of this timer: turn on the red light for the first 15 seconds of the cycle (while the accumulated value is less than or equal to 15; then turn on the Green light for 10 seconds (while the accumulated value is greater that 15, but less than or equal to 25); finally, turn on the Yellow light for 5 seconds (while the accumulated value is greater than 25 and until the timer is done at 30 seconds and resets itself, initiating the cycle again).

D. Math Instructions:

Now we will explore how to use the abovementioned mathematical capabilities of our friend PLC to actually perform mathematical operations based on quantities stored on data memory word registers (or connected directly to Analog Inputs), based on the logic to the left of these instructions to become true, and store the result in a destination data memory word registers (or connected directly to Analog Outputs).

1. Add, ADD : When the logic to its left becomes true, the ADD instruction adds together the values from Source A and Source B and stores the result in the destination register (Destination). Again, care should be taken with add as well as with all mathematical instructions not to get unexpected logic results if we mix different data types in Source A, Source B, and Destination. ADD is an output instruction.

ADD	
Add	
Source A	N7:0
Source B	N7:10
Dest	N7:20

2. Subtract, SUB : When the logic to its left becomes true, the Sub instruction subtracts the quantity in Source B from the quantity in Source A, and stores the result in Destination. SUB is an output instruction.

SUB	
Subtract	
Source A	N7:5
Source B	N7:15
Dest	N7:22

3. Multiply, MUL : When the logic to its left becomes true, the MUL instruction multiplies Source A times Source B and stores the result in Destination. Care should be taken not to allow the logic to try to place a result greater than the Destination data type boundary (for example, greater than 32,767 for a 15 bit Integer), as we can get unexpected operation or a PLC fault. MUL is an output instruction.

MUL	
Multiply	
Source A	N7:7
Source B	N7:14
Dest	N7:21

4. Divide, DIV : When the logic to its left becomes true, the DIV instruction divides Source A between Source B, and places the result in Destination. Here, a good programming practice is to accompany the DIV instruction with further logic, placed before the instruction execution that will prevent Source B to become 0. If we try to divide by zero, we get unexpected logic results, and in some cases, the PLC goes to fault. DIV is an output instruction.

DIV	
Divide	
Source A	N7:8
Source B	N7:16
Dest	N7:24

5. Absolute Value, ABS : When the logic to its left becomes true, the ABS instruction determines the Absolute Value of Source and places it in Destination. ABS is an output instruction.

ABS	
Absolute Value	
Source	N7:0
Dest	N7:5

6. Sine, SIN : When the logic to its left becomes true, the SIN instruction assumes that Source is in radians, determines its Sine trigonometric function, and stores the result in Destination. If we happen to have a quantity in degrees, we will discuss now two more instructions that can help us do the correct trick. SIN is an output instruction.

SIN	
Sine	
Source	N7:0
Dest	N7:15

7. Degrees to Radians, RAD : When the logic to its left becomes true, the RAD instruction assumes that Source is in degrees, converts it to radians, and stores the result in Destination. RAD is an output instruction.

RAD
Degrees to Radians
Source N7:0
Dest N7:15

8. Radians to Degrees, DEG : Conversely to the RAD instruction, the DEG instruction assumes that Source is in radians, and, upon the logic to its left becoming true, converts it to degrees, and stores the result in Destination. DEG is an output instruction.

DEG
Radians to Degrees
Source N7:0
Dest N7:17

9. Cosine, COS : When the logic to its left becomes true, the COS instruction assumes that Source is in radians, determines its Cosine trigonometric function, and stores the result in Destination. COS is an output instruction.

COS
Cosine
Source N7:3
Dest N7:15

10. Tangent, TAN : When the logic to its left becomes true, the TAN instruction assumes that Source is in radians, determines its Tangent trigonometric function, and stores the result in Destination. TAN is an output instruction.

TAN
Tangent
Source N7:4
Dest N7:15

11. Natural Logarithm, LN : When the logic to its left becomes true, the LN instruction takes the natural logarithm (basis e, or 2.71828) of Source and store it in Destination. LN is an output instruction.

LN
Natural Log
Source N7:0
Dest N7:10

12. Logarithm, LOG : When the logic to its left becomes true, the LOG instruction takes the logarithm (basis 10) of Source and stores it in Destination. LOG is an output instruction.

LOG	
Log Base 10	
Source	N7:2
Dest	N7:10

13. Square Root, SQR : When the logic to its left becomes true, the SQR instruction extracts the square root of Source and stores it in Destination. SQR is an output instruction.

SQR	
Square Root	
Source	N7:4
Dest	N7:15

14. X to the power of Y, XPY : When the logic to its left becomes true, the XPY instruction raises Source A to the power of Source B (Source $A^{\text{Source B}}$), and stores the result in Destination. Again, as mentioned in the discussion for the MUL instruction, care should be taken not to allow the logic to try to place a result greater than the Destination data type boundary. XPY is an output instruction.

XPY	
X To the Power of Y	
Source A	N7:0
Source B	N7:10
Dest	N7:20

15. Negate, NEG : When the logic to its left becomes true, the NEG instruction determines the negative of Source (or multiplies it by -1), and stores the result in Destination. NEG is an output instruction.

NEG	
Negate	
Source	F8:4
Dest	F8:10

16. Limit Test, LIM : This instruction evaluates to True if the quantity in Test is greater than or equal to the quantity in Low Lim and less than or equal to the quantity in High Lim (Test is between Low Lim and High Lim); otherwise evaluates to False. LIM is an input instruction.

LIM	
Limit Test	
Low Lim	N7:0
Test	N7:10
High Lim	N7:20

17. Compute, CPT : Let's review now the case where we need to perform a more "complex" mathematical operation, which is either not covered by the presented instructions, or that will result in too long or impractical sequence of instructions and memory buffers to store intermediate calculations. Some PLC processors allow us to write mathematical expressions and execute them with the Compute instruction. When the logic to its left becomes true, the CPT instruction evaluates the Expression, solves it, and stores the result in Destination. Use the operators +, -, *, /, ** for sum, subtract, multiply, divide and raise to a power exponent, respectively, and use parentheses pairs to determine the order of mathematical operations. Be careful when applying multiple parentheses pairs, as if they are not complete, the PLC will not be able to compile the instruction, and will give us an error when trying to add it into the code. CPT is an output instruction.

CPT	
Compute	
Dest	N7:20
Expr	
(N7:0 + N7:1) * 5	

E. Bitwise Logical Instructions:

We now go ahead and learn how to perform logical operations in words, but in a bit-by-bit fashion.

1. Bitwise And, AND : When the logic to its left becomes true, this instruction performs a logical AND operation on a bit-by-bit basis for the words in Source A and Source B, and stores the result in Destination. The results table for the logical AND operations follows, as well as an example of how does the AND instruction operates:

AND	
Bitwise AND	
Source A	B3:5
Source B	B3:7
Dest	B3:10

AND	0	1
0	0	0
1	0	1

Take the value of the bit in the column, and look for the cell in the intersection of the value of the bit in the row. This cell will have the value of the result of the AND operation between these two values. For example, 0 AND 0 = 0; 1 AND 0 = 0; 0 AND 1 = 0, and 1 AND 1 = 1. The logical AND operation is the Boolean logic simplest representation of two switches in series connection: for the arrangement to allow the passage of current (have 1 as a result), both switches should be on (both bits should be 1)

Now, let's evaluate an example of the operation of the bitwise AND instruction. In order to simplify things, we will assume for this example that words only consist of 8 bits, numbered from 0 to 7:

↓Word/Bit→	0	1	2	3	4	5	6	7
Source A	0	0	1	1	0	1	0	1
Source B	0	1	0	1	0	0	1	1
Destination	0	0	0	1	0	0	0	1

AND is an output instruction.

2. Bitwise Or, OR : When the logic to its left becomes true, this instruction performs a logical OR operation on a bit-by-bit basis for the words in Source A and Source B, and stores the result in Destination. The results table for the logical OR operations follows, as well as an example of how does the OR instruction operates:

OR	
Bitwise OR	
Source A	B3:0
Source B	B3:10
Dest	B3:20

OR	0	1
0	0	1
1	1	1

For example, 0 OR 0 = 0; 1 OR 0 = 1; 0 OR 1 = 1, and 1 OR 1 = 1. The logical OR operation is the Boolean logic simplest representation of two switches in parallel connection: for the arrangement to allow the passage of current (have 1 as a result), any switch should be on (any bit should be 1).

Now, let's evaluate an example of the operation of the bitwise OR instruction. In order to simplify things, we will again assume for this example that words only consist of 8 bits, numbered from 0 to 7:

↓Word/Bit→	0	1	2	3	4	5	6	7
Source A	0	0	1	1	0	1	0	1
Source B	0	1	0	1	0	0	1	1
Destination	0	1	1	1	0	1	1	1

OR is an output instruction.

3. Bitwise Exclusive OR, XOR : When the logic to its left becomes true, this instruction performs a logical XOR operation on a bit-by-bit basis for the words in Source A and Source B, and stores the result in Destination. The results table for the logical XOR operations follows, as well as an example of how does the XOR instruction operates:

XOR	
Bitwise Exclusive OR	
Source A	B3:0
Source B	B3:10
Dest	B3:20

XOR	0	1
0	0	1
1	1	0

For example, 0 XOR 0 = 0; 1 XOR 0 = 1; 0 XOR 1 = 1, and 1 XOR 1 = 0.

Following our established pattern, let's evaluate an example of the operation of the bitwise XOR instruction. In order to simplify things, we will assume another time for this example that words only consist of 8 bits, numbered from 0 to 7:

↓Word/Bit→	0	1	2	3	4	5	6	7
Source A	0	0	1	1	0	1	0	1
Source B	0	1	0	1	0	0	1	1
Destination	0	1	1	0	0	1	1	0

XOR is an output instruction.

4. Bitwise Negate, NOT : When the logic to its left becomes true, this instruction performs a logical NOT operation (negates the bits) for the word in Source, and stores the result in Destination. An example of how does the NOT instruction operates follows:

NOT	
NOT	
Source	N7:4
Dest	N7:15

Again, in order to simplify things, we will assume for this example that words only consist of 8 bits, numbered from 0 to 7:

↓Word/Bit→	0	1	2	3	4	5	6	7
Source	0	0	1	1	0	1	0	1
Destination	1	1	0	0	1	0	1	0

NOT is an output instruction.

F. Data Operations

Now we will introduce a group of instructions that operate on words of data and become very handy when we try to perform some operations (like, for example, storing "live" I/O data in registers for the future use of mathematical operations):

1. Clear, CLR : When the logic to its left becomes true, CLR resets the

```
        CLR
Clear
Dest          N7:12
```

word in Destination to a value of 0. CLR is used sometimes to initialize a "buffer" or storage register at the beginning of a more complex operation, to make sure there is no number left from a prior execution. For example, if we want to add up several values and store the sum in a register, it is good to reset this register to 0 before beginning to add values. CLR is an output instruction.

2. Move, MOV : When the logic to its left becomes true, this

```
        MOV
Move
Source        N7:0
Dest          N7:10
```

instruction takes the value of the word in Source and stores it in Destination. Usually used to move data from one register to another or from "live" I/O data to a register. MOV is an output instruction.

3. File Copy, COP : Now we will introduce the concept of a word file. A

```
        COP
Copy File
Source       #N7:0
Dest         #N7:20
Length           5
```

word file is a group of words that have consecutive addresses inside the same data table file. It is defined by its starting address and the number of its members (sometimes called the **length** of the file). For example, if we want to group together words N7:0, N7:1, and N7:2 in order to perform a file operation, this file becomes #N7:0 with a length of 3 words. Note that the file is denoted by the "pound" sign in front of the starting address. Having said that, and having now clear what a word file is, we will say that when the logic to its left becomes true, COP takes the values of the word file in Source and copies them in the word file in Destination. For example, if we want to COP our #N7:0 file to #N7:100 with Len of 3 words, the value in N7:0 will be copied onto N7:100; N7:1 onto N7:101, and N7:2 onto N7:102. COP is an output instruction.

G. Program Control

We will discuss here two instructions that will be quite helpful when we get into the special topic of Structured Programming later on.

1. Jump to a Subroutine, JSR : This is an output instruction that makes the logic execution to "jump" from one Program File to another, when the logic to its left becomes true. This allows us to execute some pieces of ladder code conditionally. There is also the possibility of "nesting" subroutines, this is: jumping from the Main Program File (Program File 2) to Program File 3, and then from Program File 3, conditionally using another JSR instruction to go to another Program File 4. Then, return the execution from Program File 4 to Program File 3, to return afterwards the execution from Program File 3 to the Main Program File 2, and resume the code execution there. In order to make these returns to the calling Program File, we need to use an instruction that will be discussed next.

JSR
Jump to a Subroutine
File Number 3

2. Return, RET : This is also an output instruction that, as discussed above and as its name suggests, returns the program execution from a Program File that was called from another, to the calling Program File. The execution will then resume in the calling Program File immediately thereafter the rung with JSR instruction. Care should be taken when using JSR instructions, as the processor expects to find a RET counterpart for each one. There is a specific Processor Fault for a JSR without accompanying RET instruction.

5 SPECIAL TOPICS

We will now try to build upon the basics discussed in order to present some practical applications and illustrations of the further capabilities of our processors, as well as some topics that usually go alongside with the PLC processor. We should have in the section both a reference and a source of a better understanding of how are our friends positioned in the Industrial Automation world.

1. **Special Topic – Indirect Addressing**

We will try here to go inside a special topic which is sometimes perceived as complex, but nevertheless very useful in some programming applications. This is **Indirect Addressing**. Up to now, we have identified the PLC Data Memory addresses with a Data Type indicator (i.e. I, O, B, T, C, N, F, etc.), a Data File number, and then the element number inside that Data File, for example, N7:10. Let's now visualize that we need to refer to different element numbers inside this Data File, for example, N7:11, N7:12, and N7:13. We can create a buffer in some other memory location, again for example, N10:0. The PLC addressing allows us to use brackets and use this buffer as the memory address, say N7:[N10:0]. Now, we can programmatically place the values 11, 12 and 13 sequentially in this N10:0 buffer address, and then, the expression N7:[N10:0] will become functionally similar to N7:11, N7:12 and N7:13, respectively.

Ladder Application Example:

Consider that we need to determine the average of the last three values read from one of our analog indicators, having readings ten seconds apart.

Solution:

Initialize the Pointer by writing a value of 10 if the Pointer is empty.

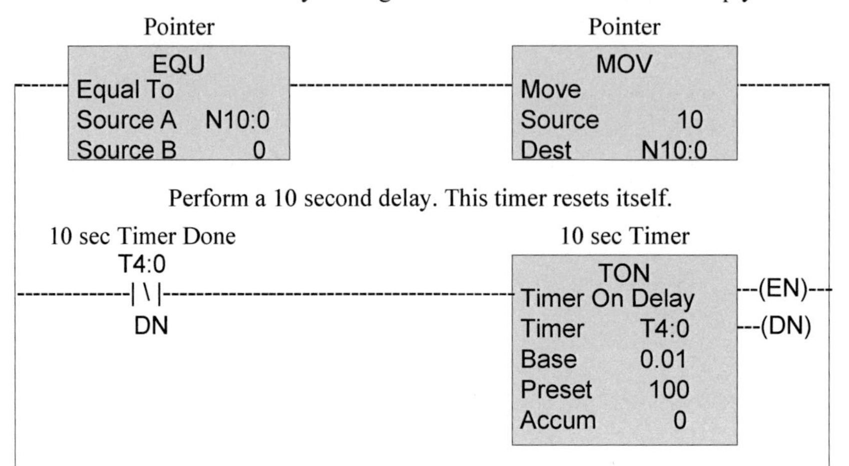

Perform a 10 second delay. This timer resets itself.

Each time the timer completes, move the Analog Input reading to the correspondent place in the readings array to be averaged. Then take the average of the three values and place into the correspondent memory area. Finally, add 1 to the pointer value to be used in the next iteration.

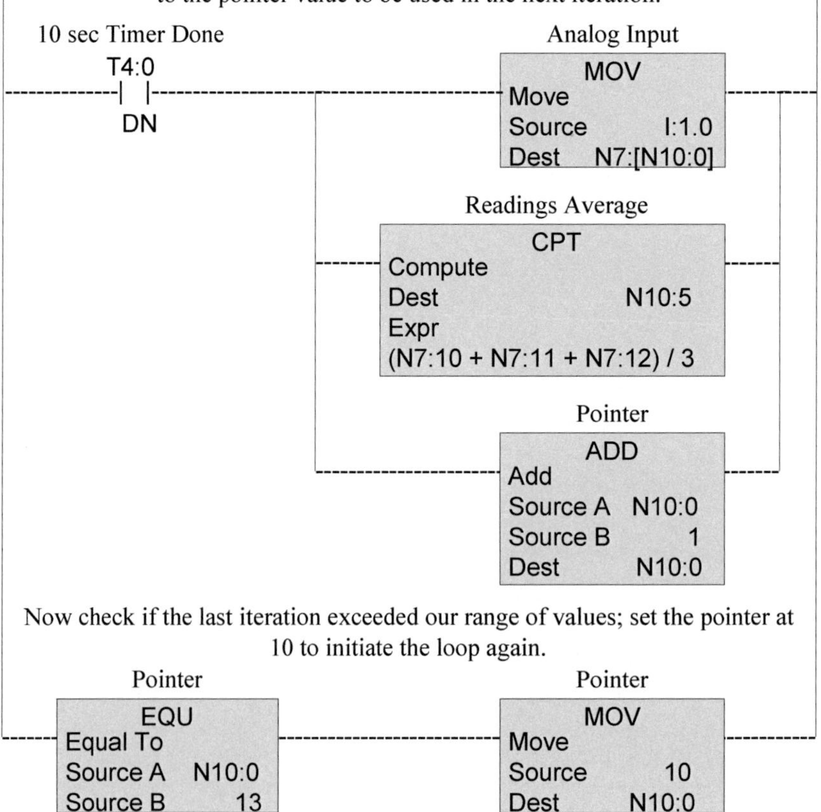

Now check if the last iteration exceeded our range of values; set the pointer at 10 to initiate the loop again.

Ladder Explanation

We first initialize our Pointer (N10:0) by writing an initial value of 10 to it when it is empty (equal to 0). Then we use again a resource we have used in many examples in this text: the self-resetting timer, in this case with a preset value of 10 seconds. Afterwards, every time the timer is done, we move the value of our Analog indicator (coming from I:1.0 – to review a little bit our addressing concepts, this is the first analog channel for a module located in the second slot of the chassis-) to the address in our array to be averaged (this is, N7:10, N7:11 and N7:12) indicated by our pointer. After this, we update our average value (to take into account the new value recently moved into the array) in N10:5, by adding together the three values in the array and dividing them by 3 (as this is a somewhat compounded, not to say "complex" operation, we use the CPT instruction rather than using 3 ADD instructions, two intermediate calculation buffers, and a DIV instruction). After these operations are completed, add 1 to our Pointer value, in order for the next iteration to discard the oldest value and put the newest value in its place. If we come out from this operation with a value of 13 (which is outside our operating range), reset the Pointer value to 10, and start the routine all over again.

2. **Special Topic – Fault Routines**

Some PLC processors allow for a ladder file to be configured as a **Fault Routine**. This is a special program that the processor will run if it detects a Fault, in order to try to recover from it. While running this routine, the processor does not declare itself in Fault (remember that, after a Fault, the processor halts operation and does not execute code in the application anymore). If after running this routine, the Fault condition persists, then the processor declares itself in Fault. In order to write code for this fault routine, we must thoroughly know and understand the code of the application residing in the processor. This knowledge will help us to identify "vulnerable areas" in the code where a Fault can be generated. For example, if we have code in which we perform a division where the dividend is a memory buffer, we should look for the possibility for this buffer to have a value of 0. The resulting attempt of division by zero will result in a processor Fault. In this case, we place code on the Fault Routine to move a value different than 0 to this buffer, so the operation can continue in a normal fashion. Also, care must be taken on the effect that this "forced" value will have in our application.

3. **Special Topic – Structured Programming**

Many international standards call for the **modularization** of the program code, this is, to try to write code in such a manner that portions of this code can be reused if facing a similar application in a different place in the program. This practice reduces, in some cases substantially, the time and effort needed to implement a particular application. The resulting code is more compact, consumes less memory and computing resources of the processor to execute, although somewhat a little harder to troubleshoot. There are some considerations that must be taken in order to develop code in this fashion: first, it is desirable to group similar devices to be operated together, and build **modules** that can use similar programming. We will consider a simple example of a motor. Let's consider the case where we have many motors with similar characteristics to perform our previously discussed start/stop routine with the use of their respective pushbuttons. We first reorder the connected I/O, or perform a "mapping" operation, where this data will be handy for the structured code to manage. Then we write the code, making use of the advantages of indirect addressing to loop through every motor needed to operate. An actual example will further explain and clarify:

Ladder Application Example:

Consider that we need to develop a program module to manage the starting/ stopping of three motors.

Perform I/O mapping from the connected I/O values to the values in the structured array.

```
      Motor 1 Start Button                                     Motor 1 Start Button
             I:0                                                      N10:1
-------------|  |-------------------------------------------------------( )-----------
             0                                                          0
      Motor 1 Stop Button                                      Motor 1 Stop Button
             I:0                                                      N10:1
-------------|  |-------------------------------------------------------( )-----------
             2                                                          1
        Motor 1 On/Off                                           Motor 1 On/Off
            N10:1                                                     O:1
-------------|  |-------------------------------------------------------( )-----------
             2                                                          0
      Motor 2 Start Button                                     Motor 2 Start Button
             I:0                                                      N10:2
-------------|  |-------------------------------------------------------( )-----------
             4                                                          0
      Motor 2 Stop Button                                      Motor 2 Stop Button
             I:0                                                      N10:2
-------------|  |-------------------------------------------------------( )-----------
             6                                                          1
        Motor 2 On/Off                                           Motor 2 On/Off
            N10:2                                                     O:1
-------------|  |-------------------------------------------------------( )-----------
             2                                                          1
      Motor 3 Start Button                                     Motor 3 Start Button
             I:0                                                      N10:3
-------------|  |-------------------------------------------------------( )-----------
             8                                                          0
      Motor 3 Stop Button                                      Motor 3 Stop Button
             I:0                                                      N10:3
-------------|  |-------------------------------------------------------( )-----------
             10                                                         1
        Motor 3 On/Off                                           Motor 3 On/Off
            N10:3                                                     O:1
-------------|  |-------------------------------------------------------( )-----------
             2                                                          2
```

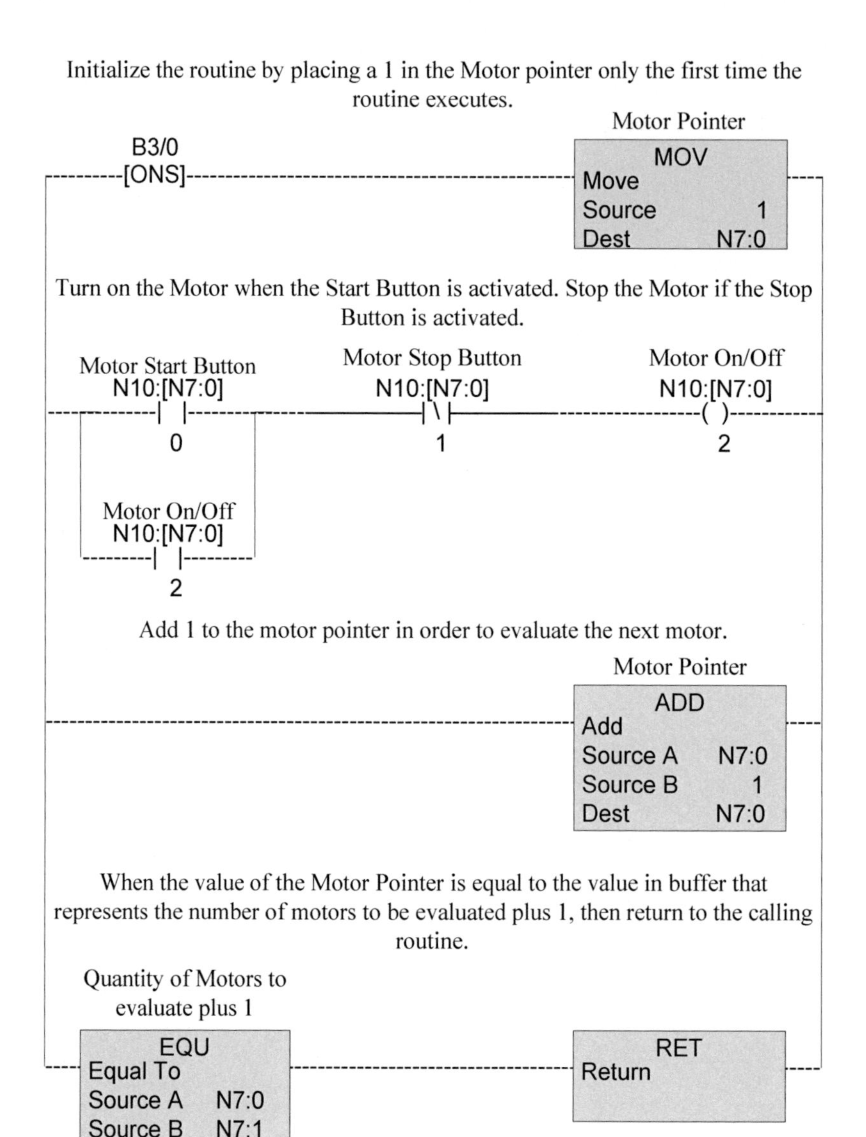

Ladder Explanation

We begin by performing I/O mapping. Note that we assign a word in file N10 for each motor, and that this word should have the same structure for every motor. In our case, we use bit 0 for the Start pushbutton, bit 1 for the Stop pushbutton, and bit 2 for the Start/Stop command. This mapping just moves the data from its connected Input address to the corresponding place in this "mapped buffer", and from the "mapped buffer" to the connected Outputs. Then, we go to the Motor Starting/Stopping routine: we initialize by placing a 1 in our Motor number buffer (N7:0). Note that there is no logic to the left of the ONS instruction, so the "logic to the left" of the ONS instruction will have a false-to-true transition each time the routine begins executing. The ONS instruction will guarantee that this initialization will happen only once at the very beginning of the execution of the routine. We then place the code reviewed before for starting and stopping the motor, with the variation that we now use indirect address to interact with our "mapped" I/O. We then add 1 to our buffer, in order to be able to evaluate the next motor, and so on, until the value of the buffer reaches the value on N7:1, a buffer where we place the number of motors to be evaluated plus 1. For the buffer in N7:0 to reach this value in N7:1 means that we are done evaluating the quantity of motors needed, so we Return the execution to the calling routine. Care must be taken, as this is a rather simple example, and in the way this code is written, can easily take care of a couple hundred devices. If we happen to go further from that magnitude, we should examine the effect that scanning this routine will have in the overall processor scan time, in order to avoid watchdog Faults. If we are to operate more than a couple hundred devices, we should consider writing logic code that will split the evaluation of the Motor quantity between more than one scan. Remember that the order of magnitude of the processor scan time (in the order of milliseconds) does not prohibit being practical if a motor is not evaluated in a particular scan. In practice, every actuation of a pushbutton will last for many (maybe hundreds of program scans).

4. Special Topic – Very Basic Troubleshooting

In order to determine the possible cause of problems or malfunctions arousing from our application, is it needed to go Online with the processor in Run mode. Most programming software packages provide means to visually identify that the processor is in Run mode (the first obvious cause for a problem will be the processor in Program mode, where the code is not executed, thus not operating any Outputs), besides somehow identifying instructions that are evaluating true. We will use for example purposes a software that highlights the instructions that are evaluating to true, as well as highlighting the power rails (the vertical lines alongside the rungs, or the "poles" of the ladder) to indicate Run mode.

Begin with the de-energized output, and work back to identify one or more de-energized instructions. If the instruction refers to an internal memory buffer (not I/O), check other instances of the usage of that same address in the code. Most programming software packages include some means of **cross-referencing**, where we can obtain a listing of the occurrences of the usage of that address throughout the code. For the case of a memory buffer, seek where this address is energized or de-energized (OTU, OTE, or OTL instructions; CLR or Math instructions referring to the whole word containing the address) and try to work back to the cause of the instruction evaluating to false. In the case of actual I/O, maybe electrical measurements are needed to identify if there is power energy at a discrete I/O connection point, or the presence of an analog signal (most commonly 4-20mADC) at an analog I/O connection point.

Consider also (by making use of the cross-reference capability) if there are duplicate instances of OTE instructions referencing the same address. This causes unpredictable operation of the processor, regardless of the logic attached to each instance. If you need to energize the same address in different places on the code, rather use a different auxiliary bit (bits in the B3 File can do the trick) to be energized in every case, and then parallel all of these bits and OTE the needed address, resulting in one and only one OTE instruction for the address. This principle does apply only for OTE instructions, not for OTL or OTU instructions.

Now we present a visual example of the Online editor screen:

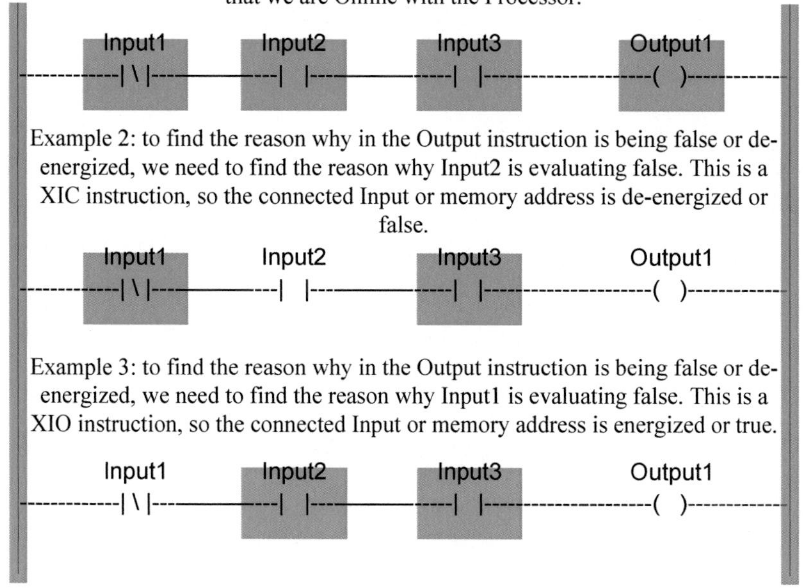

5. Special Topic – PLC Processor Battery

Most processors draw their power to operate from the chassis power supply. This very same power source maintains the processor memory, both the data and the program files. As most processors base their storage in volatile memory (which is erased should the power applied to it fails), they need a backup power source to maintain the program and data integrity should the power source fail. This is achieved by a backup battery installed in the processor module. In order to assure operational continuity, care should be taken to change this battery on a regular basis, say yearly; and, as we tend to forget such important things, duly document the date of the last battery change. Some processors offer a space with a sticker where to write down the date of the last battery change. Some newer processors are offering a slow-discharge capacitor to maintain power to the processor memory, thus eliminating the need for the battery.

6. Special Topic – PLCs vs. PACs

Although the focus of this book is to present a basic introduction to PLCs, quite often the term "PAC" (Programmable Automation Controller) is found and widely used in Industrial Automation literature and applications. In their basic and low-end (more simple) application, they are (surprisingly) actually the same. The basic set of instructions discussed here will apply (again, with minor differences mostly by manufacturer of the system) to both presentations. From here, it can be inferred that then we can, in a very simple sense, identify a PAC as a PLC with enhanced capabilities. The difference between both resides in the high-end of the PAC application. PACs are capable of applications like, for example:

- Batch Process Control (as having a product recipe, complete with commands to for example, add ingredients in the correct quantities, adding them together, mixing with the correct parameters, and transfer the mix out of a Process Unit –container-, in an ordered sequence)
- Special communications with high-end, supervisory computer applications (like Manufacturing Execution Systems, or MES, that in a simple sense, schedule and supervise manufacturing operations and gather data to present management reports)
- Multitasking (can process multiple program applications at the same time, can interrupt the execution of a piece of code to execute another under configured conditions, independently execute programs on a timely basis (i.e. every second), without interrupting the execution of others, and others of the sort.
- Acceptance of multiple programming languages, like Ladder Diagrams, Sequential Function Charts (SFCs), Structured Text, and Function Blocks. These languages are to be briefly discussed later on.

These are operations that give PACs the flexibility of performing tasks intended for more complex and integrated systems, but, again, in their low end, their basic applications rely in the same principles discussed here. Bear in mind that, in order to become knowledgeable in the programming and configuration of a PAC system, one must first be proficient on the basic functionality of the system. Detailed discussion of the high-end capabilities of PACs is beyond the scope of this text.

7. **Special Topic – Communication Networks**

We have mentioned before that one of the key attributes of our friend PLC is its flexibility. In the line of that flexibility, we have many communication options in order to go online with the processor. There are also communication options that allow the processor to manage **Remote I/O**, or chassis located physically away from the processor chassis, being this communication networks the channel that allows the processor to properly interact with that I/O. Also, as Industrial Automation technologies have advanced, we have seen the appearance of diverse **Open Systems**. These are communication-capable devices that communicate with the processor and are not necessarily pertaining to the processor manufacturer-specific I/O (usually they are not chassis-based I/O, and often not even modules). Open systems function under International standards, whose practical application is based on the creation of **Device Definition Files**. These are text-based files which, with a standard-dictated format, define the communications between the specific device and a network-specific scanner somehow attached to the processor. The beauty and increased flexibility that Open Systems bring is that we can acquire devices in the open market, not necessarily from the PLC supplier, and obtain from its manufacturer this Device Definition File (usually available for download in the manufacturer's Internet website). We then "register" this file with the scanner, connect the communications network, and both the processor and the device are now interchanging all the data that the device is capable of. Another high level application is the communication that makes possible for some PLC processors to exchange data with high-level supervisory business systems, like for example, Enterprise databases and ERP (Enterprise Resource Planning) applications. In order to provide this information as a reference, we will briefly discuss the most popular industrial communication networks serving PLC processors:

- EtherNet – there are available many industrialized versions of the very same communications network that allows the office and home computers to communicate with each other (also, the very same network that allows us to access the Internet). We use EtherNet, as previously mentioned, to communicate PLCs with supervisory systems, with other PLC processors, with programming software, and to perform the communications link to HMI (Human-Machine Interface) systems that will be discussed later in this section. Also, there exist some I/O adapters that allow entire chassis of modules to communicate with a processor. Furthermore, there are also EtherNet-capable devices, which just connect to the same kind of cabling we use for home or office computer networking, and communicate data to and from the processor, as well as many emergent technologies which make more and more use of wireless communication technology over EtherNet. Let's now visualize Industrial Automation systems in a layered fashion:

- o Having Supervisory systems, Programming software and HMI (again, we will go into HMI later in this section) at the top layer
- o Having controllers (in this case, PLC processors) in the intermediate layer, and
- o Having field instrumentation and I/O devices in general in the bottom layer

Having this layered representation of the systems under our discussion, we find EtherNet available to handle communications among the three layers of our model.

- DeviceNet- this is a dedicated industrialized network, mostly used to communicate field devices and I/O to our friend processor. This is one network that, as mentioned before, uses a scanner module in the PLC chassis in order to establish communications with the network. This scanner has a special memory which will be the means of transferring the I/O information to and from the processor. These scanners are configured and the DeviceNet device definition files, which in this case are called **Electronic Data Sheets (EDS)**, are "loaded" into the scanner memory. Now the scanner has the capability to build a "map" of the I/O data it needs to transfer to and from the processor (because an important part of the EDS file –as is the case with most device definition files- is a description of how much memory space does each device in the network interchanges with the processor, what specifically does every piece of data means, and if this data is Input or Output to the processor. Then, the next step is to try to fit this "map" in the processor data memory. For this purpose, we need to configure in the processor a memory area for the use of this scanner "map": if the DeviceNet scanner is placed in chassis slot 4, then the memory block for the DeviceNet network connected to that scanner is I:4.0 and subsequent words for Input data, as well as O:4.0 and subsequent words for Output data. For example, if we happen to use a DeviceNet-capable valve, for which we obtain the open and closed limit switches as Input data, and the Open/Close command as Output data, we configure in the scanner 8 bits (the minimum amount) of Input and 8 bits of Output. Let's say we configure the Input data to reside in the group of 8 bits beginning at I:4.2.0, and the Output data to reside in the group of 8 bits beginning at O:4.2.0 (it is not mandatory for the Input and Output images to reside in the same word number of their respective data tables, but this helps to keep some order). If the first bit in the Input image is the open limit switch and the second one corresponds to the closed limit switch, and if the first bit in the Output image is the Open/Close command, then we will use address I:4.2.0 for the open limit switch, I:4.2.1 for the closed limit switch, and finally O:4.2.0 for the Open/Close command. We find DeviceNet servicing the interface between the intermediate and bottom layers of our systems representation.

- ControlNet – this is another dedicated industrialized network, which serves mostly the communication between processors and between the processor and remote I/O chassis. It has a very reliable performance. We can find ControlNet often used in the intermediate layer of our layered systems representation.

- Foundation Fieldbus, Profibus – We group these two together as they have very similar characteristics. These are industrialized networks which primary use is to interact with specialized field instrumentation (for example, but not limited to: temperature, pressure and flow transmitters, as well as control valves –we can make the analogy while being correct most of the time, that these two networks interact with devices that we have discussed before as being Analog I/O). The difference and the beauty of these networks show up when we compare them to "traditional" Analog I/O. When we have our "traditional" I/O, there is a single analog cable that transmits a signal (quite often 4-20mADC) analog to the magnitude of the signal, as previously discussed. For these networks, rather that a milliamp or millivolt signal, very often through the very same cable, digital pulses carrying larger amounts of information (among them the same analog signal, along with diagnostic data and others) are transmitted. Besides that, a single cable can be connected to adapters (sometimes called **spur adapters**) that service many field instruments. Of course the field instruments should be capable of this communication. We find these networks servicing the interface between the intermediate and bottom layers of our systems representation.

- Modbus – this is a very popular network, initially developed by one PLC manufacturer, but that has made its way to become an Open System as well. Here, instead of a Device Definition file, Modbus has a standard data allocation, in what the standard calls **Coils** (for discrete memory locations) and **Registers** (for analog memory locations). If we acquire a Modbus-capable device (and there is a considerable quantity in the market, from field devices, PLC systems, up to electrical power meters, among its wide variety), its documentation (i.e. manuals) will tell us the data available for communications, and in which Coil or Register numbers it is located. With this information in hand, our Modbus scanner can move data to and from this registers to memory addresses in our processor (or for some processors to use it directly as native data addresses). We find Modbus servicing the interface between the intermediate and bottom layers of our systems representation.

- Serial Communications – we will discuss here a wide variety of communication arrangements, all of them based on legacy serial protocols, but still very often used nowadays. All of these communications arrangements (note that we are not using the term "networks", as most of these arrangements can only work in a point-to-point configuration servicing two devices) are somehow based in serial communications protocols, being RS-232 and RS-485 the most popular. Some PLC manufacturers have developed their own proprietary variants, which sometimes are used for inter-processor communications. These arrangements often use our computer serial port (or most recently, our USB port), and connect directly to the device. Many PLC processors offer a serial port in the processor itself, often configurable to more than one protocol (but one at a time). Some PLC systems (specially the smaller processors) use serial as their (sometimes their only) means to communicate with the programming software. So we will find ourselves often in the situation where we have our computer connected with a cable (sometimes standard, off-the-shelf, electronics-stores-available; sometimes proprietary) directly to a serial port in our processor in order to perform programming. There is also the case in smaller systems where serial communications (again with the choices of cabling described above) are used to perform HMI (I am not sure if I have mentioned before that we will go into HMI later in this section) communications. In summary, their use is very specific but very common, and we find these arrangements servicing specific applications in the top and intermediate layers of our representation. In some less frequent situations, we can find applications servicing the interface between the intermediate and bottom layers of our representation.

8. **Special Topic – Other Programming Languages**

We have been discussing the ladder diagram as the principal programming language tool for programming, configuring, and troubleshooting our processors, but it is worth mentioning that some systems (specially larger ones and more often PACs) give us the flexibility of allowing programming in other languages, which we will briefly describe:

- Function Block Diagramming (FBD) – this was originally the language of choice for larger, mission specific processors called **Distributed Control Systems** (**DCS**, as opposed to PLCs). It consists on pages of diagrams composed of Function Blocks, each one encomprising different (and sometimes combined –here is why some people prefer them for the alleged ease of programming- capabilities) functionalities. For example, we can find a block that takes care by itself of the starting/stopping logic of a motor or device, being that for that same service we need at least a full rung on our ladder logic). In the Function Block diagram, we only need to "connect" the graphical representation of our Inputs, Outputs, and the auxiliaries (cascades, permissives, etc.) to the device controlling block,

Function Block Diagram Example

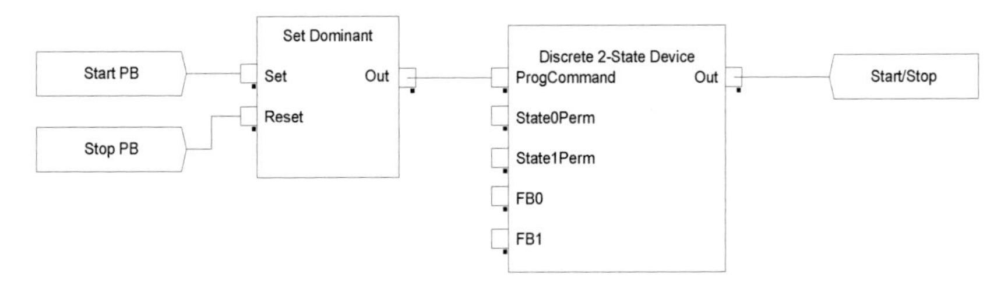

and the Block takes care of performing the corresponding logic. On the other hand, for the people that finds ladder way easier and convenient, there are some applications, like for example, latching a discrete memory address, that involve the use of rather more complex Blocks in FBD. The truth is that there are applications that are more suited for one language or the other. Generally, continuous process and analog control applications are regarded as best suited to be controlled by FBD, while discrete, high speed manufacturing applications are for Ladder. Each PLC, PAC, or DCS supplier has its own set of Function Blocks with sometimes similar, sometimes specific capabilities. We do need, the same we did with our Instruction Set section, to become familiar with the available choices and functionality for each Block before attempting perform programming in FBD. Some PACs allow the flexibility to have FBD pages as well as Ladder files (as well as some other languages to be discussed) coexisting in the same application.

- Sequential Function Charts – this is usually a feature of larger processor systems or PACs. SFCs consist of arrangements of blocks (each composed of a procedure, usually a Ladder file or FBD page) that operate in a specific sequence, with configured transitions so that each block (and its underlying ladder code) executes in order and when specific (of course, configurable) conditions exist. We can visualize the SFC as a supervisory software for the PLC: we write our ladder code for different applications in different Ladder files, and then write this special SFC code to make these applications to execute in a predetermined sequence, sometimes even in parallel. SFCs have applications in batch processing. Let's take for example an industrial bread making operation: we can write individual ladder files for the operations we need to perform. We write a file for loading the flour, another for charging water, another for charging salt and species (we get fancy here with our bread), another for mixing, another to charge the mix into the ovens, and a last one to discharge the baked product. Then we develop a SFC where these operations (one in each SFC block) are called in order, conditioning the execution of each one on the previous one to be completed (i.e. won't mix until the ingredients are completely charged). We then have the added convenience that only the code needed for the operation that is being performed is executed, and that we can, while monitoring our SFC on line, have a higher-level, easier view of how our process is doing, rather than looking through the ladder code.

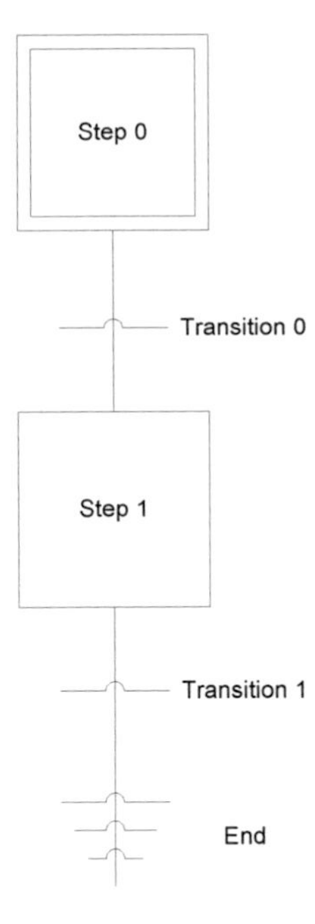

- Structured Text – as its name implies, this is a completely text-based language, where every ladder instruction as well as branching has its own mnemonic to be "translated" into text in order to build program statements with them. It is more suited for those who are more familiar with other traditional "text-based" programming languages like C, Visual Basic, and others.

9. **Special Topic – Human-Machine Interface (HMI)**

At last we get into the subject that we have been mentioning and "announcing" throughout the text. Up to the moment, we have been discussing how to interact with our friend PLC for configuration, programming and troubleshooting using the programming software. There is a potential problem here: even though we have seen the "virtues" of these software packages, it is not always of practical usefulness that everybody who needs access to the information in the PLC processor will access it by using the programming software. For example, in many applications, we do not want for the process operators to have access to the PLC processor in some fashion that they can modify the code. But in order to properly operate and interact with their process, they nevertheless need access to important status information, as well as to write some process setpoints and commands. In order to grant this processor access to non-programmers or non-maintenance personnel, various software packages and/or devices have been developed to allow an "insight" into the status data available at the processor, as well as write access to predefined command and setpoint buffers. As these applications allow the human being to interface with the process (or machinery) being controlled by the processor, these applications are named **Human-Machine Interface**, or **HMI**. This is specialized software that has read and write access to the PLC processor data memory. This software can reside and run in a standard computer, or it can be a specialized terminal (sometimes called **Operator Interface Terminal** or **OIT**). We have already discussed the communication means between these devices and the PLC processor. There are usually color graphic screens (for smaller –and less expensive- systems we can also find monochrome terminals, and even text-only terminals).

In order to grant process visibility, each application has its own (but all quite similar) means of presenting data: mostly, analog data is presented in the form of numeric indications in the screen, and digital data is presented in two principal (among other) ways: the ability for a screen object to change its color based on the state of the read digital data bit, or the visibility or invisibility (to appear or disappear from the screen) of an object. Of course, every software has its own, many "fancies", like, for example "filling" bargraphs that give us the visual sense on an analog variable level, and many other of the sort, but in most applications, we will somehow see the numeric indicators, the color-changing objects, and the appearing/disappearing objects. In order to allow commands to be performed, you can "**animate**" (which is the term used to define a screen object that is configured some dynamic property depending on read data changing in the PLC processor memory) an object to write a certain value to the processor memory when it is activated by a mouse click; there are also "numeric entry fields" which can be associated by processor memory addresses, and where the figure entered in downloaded into the processor memory address. Among the "fancies", we can mention the "sliderules", which download a value into the processor address depending on the position where the sliderule is "dragged" with the mouse, among others.

Usually, HMI applications are based in **tags**, which represent the different pieces of data to be either read or written. This leads us to the first configuration element, which is the **tag database**. Here, we configure which elements are to interact with the PLC processor memory, and in which processor address they can be found. So, visualize the database as the communications "roadmap" between the HMI and the PLC. When we have our tag database ready, we can now go ahead and begin drawing and animating objects in the screen, usually inside the framework of a "drawing" or graphical representation of the process flow diagram. This makes the whole HMI application more intuitive to the operator. To go again to the "fancy" limit, we can find packages that give us the ability to embed process area pictures, and then in turn superpose the animated objects over the picture.

It is worthwhile to mention a special feature often used as a communications channel between HMI systems and PLC processors. Let's imagine the complexity of the case for a developer of HMI software who intends for its system to be open (compatible with as many processors as possible): we run into the limitation that every processor has its own communication protocol, resulting in many of them. To overcome this "language" barrier, special software applications have been created, capable of communicating with the processor in its own terms, and exposing the information to the MS Windows (Note of the author: I don't intend to introduce any preference here: only I am mentioning what I have experience with and obviously know that exists. I am not sure of the existence of the same for Linux, Mac, or other OSs, therefore the specific mention) operating system, therefore read and written there by the HMI software. These special applications make use of a feature called **OLE (Object Linking and Embedding) for Process Control, or OPC**. We can nowadays find OPC packages for almost any processor in the market.

10. **Special Topic – Software Simulation**

We would like to spend some time discussing a special tool which becomes very handy when testing developed code. Some processor suppliers include as a part of their product offering software packages that allow the computer to mimic the functionality of the PLC processor. It is possible to establish this "virtual" processor, download code into it, and place it in Run mode **without interaction with any I/O**. If we happen to be able to somewhat substitute the operation of our Inputs, we can monitor the operation of our ladder code in an almost real environment, having the characteristics and the processing speed of an actual processor. Sometimes it is imperative to test the code at something near the actual processing speed and conditions without taking the risk of operating actual I/O. In these conditions, the simulation software becomes, as mentioned, very practical.

11. Special Topic – PC Based Controllers

We discuss this topic along with the previous one quite altogether, as some confusion can arouse between them. Another software product among the offering of some PLC suppliers is another special software package that, as in the Simulator case, allows the computer to mimic the functionality of the PLC processor, but in this case with the ability to operate actual I/O. This product has to function coupled with a communication adapter which can allow the computer to communicate with I/O. This is not difficult if we take a look at the variety of the discussed communication networks that service I/O, many of them with interfaces residing in the computer.

12. Special Topic – I/O Forcing

I/O Forcing is a special troubleshooting and code debugging feature of most processors. It consists on giving us the ability to fix (or "force") the value or the status of either Inputs or Outputs on our processor. For example, we can simulate or "force" a Digital Input to be energized regardless of the status of its physical connection. We can also energize a Digital Output, again regardless of the logic code that manages it. Again, there are differences upon the extent of the I/O forcing capabilities depending on the processor types. Generally, discrete I/O forcing is available in most processors, while analog I/O forcing is limited to newer processors or PACs. Forcing is always a delicate issue which need the greater amount of care in the PLC coder's or maintainer's mind, because there is no programmatic way of manipulate an output once forced, and also the program would not respond to any changes in an input if also forced. This sometimes poses a personnel and equipment safety risk, as the process interacting with the PLC will not necessarily behave as expected. For example, an Emergency Stop button will not be able to stop a motor whose connected output is forced. Care must be taken to, when troubleshooting, testing, or debugging is completed, remove and/or disable all enabled I/O forcing. Most processors have indicators, both in the Programming Software and in the processor physical module, telling us that I/O Forces are present.

13. Special Topic - A very basic primer in Automation project development

This special topic does not pretend by far to completely cover this very important and complex subject, but rather to identify some major points to consider. They apply the very same, regardless of the project size. Our discussion will focus and be limited to PLC and HMI code development. The first and most important is to always have a clear and documented scope of the job at hand, which will be the roadmap to follow during the project development. From the scope documentation (that can be as simple as the minutes of an informal meeting with those in charge of the project all the way up to a qualification document for a regulated industry), it should be possible to obtain a listing of the devices to be sensed and managed (in other words, all the **conditions** to be looked at as well as the **actions** to be performed). This listing will become our I/O list. This list will dictate the **size** of our project, will determine **which I/O modules** do we need, will help us to determine **which communication networks**, if any, are to be used, will guide us on **how many processors** we need for the task, and (after determining which I/O modules to use) will guide us to develop the **wiring diagrams** (the drawings that will depict how our system is to be connected to the outside world) (not that bad for a single list, isn't it? – these many developments from this single list should give us a hint that this listing is the heart of our design). After the I/O list is complete and the communication networks as well as the I/O modules are determined, we proceed to **set the I/O address** for each element in the list (as it applies to each and every step of the project, document, document, document, and document. Well documented I/O descriptions, code, drawings and other deliverables are often an indication of a clean and of a good quality job).

Then and only then we can look at the specifications in the scope documentation and start to develop code. Pay attention to the possibility of using structured programming, and to the consistency of your code (try to write the same code structure for the same operation in different parts of the project).

The completed code, along with the I/O listing will give us a fair idea of what do we need from our HMI. Begin by developing the tag database (again, document, document, document, and document). Then check the customer's specifications for I/O screen depictions as well as the **navigation** between screens (how or with which commands we go from one screen to another, and which is their hierarchy). Usually, the process screens follow the process diagrams (in many cases, the Process and Instrument Diagrams, or **P&IDs**). Pay attention to the quality of your HMI screens (remember that this is the element of your design that will by far be more looked at during the functional life of your project). Avoid cluttering (do not "overfill" the screen with information; do not put too many information too close to each other element), try to establish a visual balance on your screen (that the left part of the screen has more information than the right part, or the top or the bottom; that the information is not cluttered at the center of the screen and the borders are "empty", etc.), and always try to use colors that are friendly to the eye and will not cause fatigue (remember that most probably, your screens are to be looked at by Plant operators by extended periods of time).

More often than not, it becomes quite healthy to have a documented agreement from the customer stating that the project deliverables are those specified, and that the Automation projects works as intended and does what it is supposed to do. In order to be able to have such a document, we need to design and agree with the customer to some testing, either at the developer's site (Factory Acceptance Test) or after installed at the customer's facility (Site Acceptance Test) or both.

14. **Special Topic – Reporting**

If our PLC projects will be used to run important industrial processes (as it is the aspiration of most of us), it is very probable that historical (past) data (both digital and analog), as well as trends and reports from this data will be needed for process analysis and documentation. It is the worthy to mention here briefly that there are software packages (of a myriad of sizes and capabilities) which purpose is to "read" data from our PLC system. Then this data is stored in some kind of secure database in order for this data to be retrieved at a later date for the abovementioned purposes. Usually these software packages include the ability to connect to the PLC network (top level of our network model), and extract the data via OPC. Care must be exercised in the network design to bear in mind the additional loading that the Reporting activity will bring to our networks.

ABOUT THE AUTHOR

Elvin Pérez Adrover is from Puerto Rico (Born in Ponce, raised in Adjuntas, currently living in Juncos). He has about 20 years Automation experience in various Industries; Portland Cement manufacturing, Biotech, Pharmaceutical, and Wastewater Treatment, to name some. He has also been involved in Industrial Automation equipment and software sales. His background is in Chemical Engineering. He is an avid Jeep® vehicle enthusiast and philatelist. He can be reached at elvperez@coqui.net